A Sacred Trust

A Sacred Trust
Ecology and Spiritual Vision

❧

Edited by
DAVID CADMAN
&
JOHN CAREY

THE TEMENOS ACADEMY
&
THE PRINCE'S FOUNDATION

TEMENOS ACADEMY PAPERS NO. 17

First published 2002 by
The Temenos Academy
19–22 Charlotte Road
London EC2A 3SG, UK

Registered Charity No 1043015

ISBN   0 9540311 0 5 *cased*
ISBN   0 9540311 1 3 *paper*

Typeset by Colin Etheridge
Printed by Smith Settle
on Five Seasons
recycled paper.

Contents

❧

HRH THE PRINCE OF WALES
Preface

❧

It seems to be becoming harder and harder in this age to stick to what we believe—or feel. We are told constantly that we have to live in 'the real world'—but 'the real world' is *within* us. The reality is that 'Truth, Goodness and Beauty' in the outer, manifested world are only made possible through the inner, invisible pattern—the unmanifested archetype.

Occasionally I pluck up enough courage to speak about the Sacred. It is interesting that when I do a strange, palpable silence descends upon the audience. You can tell that there is a subtle, harmonious chord that links us all on a deeper level and yet most of us are terrified to admit to feelings that might be considered 'odd' or, indeed, slightly mad. But it is those 'feelings'—call them 'intuitive promptings' perhaps—which ultimately produce order out of chaos, civilized values rather than barbarism, revealed Truth instead of misguided, arrogant eccentricity.

The great perennial wisdom of the ancients reminds us over and over again that the key to the human dilemma lies in the subjugation of the ego and the search for the impossibly difficult goal of *humility*. All the traditional arts stress the need for humility and, above all, for harmony with the pulsating heart of the Cosmos— whether in Architecture, Music, Literature or Art. The same applies to Agriculture, Medicine and Education where the need for a rediscovery of Tradition has never been more urgent

During the twentieth century, the archetypal child in us was thrown out with the bath-water. My life so far has been dedicated to re-integrating that archetype; that child-like humility which is all that can keep us in touch with our otherwise forgotten imagination. I believe that the challenge of the new century lies in the urgency with which we address the need to re-instate that abused archetypal pattern in our hearts so that—*somehow*—we can re-create the lost habitats of both the physical world *and* that sacred dimension which is God's mysterious gift to humanity. *Only* through this re-integration and renewal, through the re-discovery of balance and harmony that lies hidden, and waiting to be found once again, within living traditions, can we even begin to hope to meet the accumulating challenges and supreme dangers of this century and avoid what, I have come to the reluctant conclusion, are the otherwise inevitable catastrophes that await a world awash with an untold plethora of 'information' but utterly devoid of knowledge and wisdom.

I am proud to be Patron of Temenos because it keeps alight the flame of wisdom, imagination and 'irrational' mystery in an age characterized by the ever-encroaching shadows of ignorance. I was also delighted that Temenos and my Foundation collaborated to deliver the series of lectures *Ecology: A Sacred Trust* upon which this book is based. I very much hope that it will be the beginning of a deeper relationship between the two, and that this work will be carried forward to further celebrate and honour the Sacred Tradition.

DAVID CADMAN
Chairman of the Prince's Foundation
1999-2001

A Sacred Trust–An Introduction

🐦

HRH The Prince of Wales established his Foundation to connect the art of building and the making of community, bringing together both education and practical projects. In this work, and in particular in the Foundation's education programme, we have the support of the Temenos Academy, of which HRH is the patron. Temenos teaches the perennial wisdom, which has been the ground of every civilization, and provides a number of lectures throughout our academic year. It shared with us the responsibility for the series of talks *Ecology: A Sacred Trust*, upon which this collection of essays is based. The series was also supported by the Gaia Foundation.

You might ask what ecology and a sacred trust have to do with the Foundation's aim to connect the art of building and the making of community. Indeed, it was the purpose of the series of talks to encourage the audience to ask such a question. But a clue to this and, indeed, to the very heart of our vision, is given by HRH in his Stephen Lawrence Lecture:

> There can be no doubt that the human condition requires not just a physical or functional sustenance, but also an emotional and spiritual nourishment, and our architecture, which is, after all, nothing more nor less than the physical form that we give to the world, serves at many levels either to enhance, or to undermine, these needs. The importance of 'place' can hardly be overstated, and it is therefore the responsibility—and privilege—of the architect to help to create locations as places of true quality and appeal.[1]

And again, and perhaps more specifically, this is evident in the contribution of HRH to the Millennium Reith Lectures when he said: 'We need, therefore, to rediscover a reverence for the natural world, irrespective of its usefulness to ourselves—to become more aware . . . of the "relationship of interdependence, interpenetration and reciprocity between God, Man and Creation".'[2]

I cite these quotations because, without them, I do not believe that we can truly understand the underlying purpose and aspiration of the Prince's Foundation. Indeed, these thoughts, and an insistence that mankind is a part of nature, and not apart from it, underpin and link together many of the concerns of HRH. Architecture and the built environment, organic farming, integrated medicine, the dangers of genetic modification, and the future of the countryside: all of these come within and are informed by the compass of this vision. It is no surprise, therefore, that these same thoughts and concerns lie at the heart of the Foundation.

With this, then, as a background, and as Chairman of the Foundation, it was my intention that the series of lectures, and now this collection of essays, should explore the values that underlie our present concern at the degradation of the web of life. For it seems to me that in this our dilemma is not primarily one of technology, nor even of government (important though both of these are), but rather it is a dilemma of values—at root, indeed, it has a spiritual dimension.

The essays explore this dilemma. Three of them, that of the late Philip Sherrard and those of Suheil Bushrui and Seyyed Hossein Nasr, are taken from earlier Temenos lectures; and each of the essays comes with its own voice. To try to summarize them would be to deny their intended diversity. Indeed, we see this diversity as a strength, and have deliberately resisted the temptation to overlay it with any editorial uniformity. We hope that the essays will stimulate others to explore this holy ground.

This publication has been made possible by a generous bequest from the estate of the late Violet Susan Pearson: a poet, a weaver and a countrywoman, who lived her life in the Valley of the Lune, near

Kirkby Lonsdale in the County of Cumbria. Susan Pearson was an admirer of Kathleen Raine and of Temenos from its earliest days. She wished very much that this bequest should go towards the work of Temenos, and we are most grateful to her for her support.

PHILIP SHERRARD
For Every Thing That Lives is Holy[3]

Although we may not know quite where the odds now lie in what may be called the battle for survival, of one thing at least we can be sure: that unless we reverse the premises of the type of thought and action productive of our present techno-scientific inferno, we will not escape the disaster towards which it is ineluctably propelling us; and in this paper I want to bring into focus the nature of the main premise that has to be reversed, and to put it in the context of a drama which is not to do simply with a phase of our local European history, but is archetypal in the sense that it is intimately bound up with the whole ambiguity of human existence and the whole dilemma of human destiny as envisaged in the Christian tradition. It is a drama, that is to say, which has to do with both our fall and our resurrection; for the reversing of the premises of the type of thought lying behind our present plight entails no less than the reversing of a process of ignorance, through which the distortion of our capacity to perceive the reality of things leads to our enslavement to an illusory world entirely of our own invention; and the reversing of that process is simultaneously the prelude to our regeneration, and such regeneration is simultaneously a return to a state of being and consciousness which can only be described as paradisal.

I referred just now to the distortion of our capacity to perceive things truly. What do I mean by this? The answer to that question pitches us into the centre of the arena, because it leads us directly to defining this most pernicious of the premises we are called upon to reverse. For behind this distortion lies our virtually unquestioned acceptance of the belief that as we see things so they are, or that the way in which we perceive things with our ordinary consciousness corresponds to the reality of these things—a belief which we encapsulate in the phrase 'seeing is believing'. And behind this lies in its turn something more sinister. Behind it lies a particular mental outlook, an outlook implicit in such statements as that made by Hamlet, that 'There is nothing either good or bad but thinking makes it so', or as the Cartesian 'Cogito ergo sum'—statements which, as Shakespeare was well aware but as Descartes appears not to have been, subsume the distortion about which I am talking. For what is asserted in them is not simply the notion that human thought is the determining factor of all things, including our own existence; but also that this thought is capable of providing us with a valid type of knowledge. And it is, finally, behind this notion that there is or can be a valid type of purely human knowledge that lies the premise to which I have been referring.

I will be more explicit. There are two factors that we have to grasp if we are to escape from the process of ignorance in which we are entrapped. The first is that how we perceive things depends crucially upon the state of our consciousness, and that the state of our consciousness depends upon the state of our being. This does not mean that the reality of the things themselves varies according to the consciousness which perceives them, and still less that their existence is dependent upon their being perceived. It simply means that how they appear to us, the kind of reality we attribute to them, and whether we see them as they are or, as it were, through a distorting lens, have very little to do with the things themselves and very much to do with the quality of our own being, the purity of our soul and the level of our intelligence. And this in its turn means that the way in which we see things may not correspond in the least to the reality of the things themselves. If our consciousness is dominated by a host of illusory ideas, then how we perceive things will be correspondingly illusory. And the fact that the great majority of mankind at a particular period may perceive things in a certain way does not in the least alter this: the mass of mankind may simply be enslaved to a particular set of delusions, and its perception will be conditioned accordingly.

In other words, what we perceive by means of the senses, and how we perceive it, as well as the manner in which we investigate it, are always conditioned to conform to the hidden systems of action and reaction, belief and thought, which at any particular time happen to dominate our consciousness. It is the prevailing conceptual paradigm of our consciousness, and the reality we attribute to it, that determine what we think is real and what we think is unreal. It is this paradigm, in which we believe often without being aware that we believe in it, that constitutes for us the ultimate reference point or touchstone according to which we distinguish between what we regard as true and as not true, relevant or irrelevant, in the data on which we base our theories and actions, and that gives them the meaning they have for us. Even what we call a 'fact', far from being self-evident, depends entirely upon a consensus of opinion among those of us who call it a fact, and this consensus depends entirely upon our common subscription to the ideas, beliefs and values built into such a paradigm. And the particular paradigm to which we subscribe will in its turn depend upon the state of our inner being and hence of our consciousness.

This is why the appeal to what is called empirical evidence—the evidence of the sense-data—is so delusory; for it assumes that our senses can perceive things in a kind of objective manner that is quite independent of our prior subscription to such a conceptual paradigm. Far from this being the case, what we think constitutes empirical evidence, let alone the way in which our senses read it, is already determined for us by our prior commitment to the presuppositions built into the paradigm to which, whether we are aware of it or not, we give our adherence. Hence not only how our senses perceive things, but also what we regard as valid empirical evidence, are entirely dependent upon the state of our inner being and our consciousness. That is why Herakleitos can say that the senses are false witnesses for people with impure souls.[4] Muddied, restless water can never reflect truly. We must always remember that we can see things only as they appear to us after passing through the filter of our own perceptual equipment, and that the degree to which this filter will admit or exclude the reality of what we see, or think we see, will depend entirely upon the modality of our own particular consciousness. And this in its turn will depend upon our state of being, on how free we are from self-deception and illusion.

The second factor that has to be grasped—and it is correlative with the first—is that how we see things with what I called our ordinary consciousness, and might better have called our untransmuted and unregenerate consciousness, is purely subjective. Such a consciousness corresponds to a state of being that is closed in on its own subjectivity, and consequently the way in which it views the world is likewise entirely subjective. On this level of things there is no objective world in the way we so often assume, and no view of the world that is objective, since there is not, and cannot be, any objective observer. When in this state, what we call our knowledge is the result of our attempt to know that which we do not know and which we think is not known. It is the product not of our knowing, but of our ignorance. It is the reflection of our not-knowing, of our non-awareness. And what we do not know and what we think is not known is the reality or the true nature of everything we think we observe and investigate. In fact, did we but possess the clarity of mind of a Socrates we would, like him, recognize that in this state the only thing we can know is that we know nothing, for in this state we are not capable of knowing anything else. Whatever else it is that we think we know is merely surmise and guess work.

Yet even to say this is to attribute too positive a status to what we think we know. For if the way in which we view things is determined by our ignorance, by our not-knowing and non-awareness, so that we cannot see things as they are in reality, we must be seeing them in a false way, in an illusory way; and consequently the knowledge that we think we have of them must also be a false, illusory kind of knowledge. And this must mean that any and every theory we may postulate about the nature of the universe, the structure of reality, or anything else, must not only be surmise and guess work; it must also and inevitably be a false theory. And this in its turn must mean that the state of our being is also in some way subject to self-deception and illusion, and that it is our own self-deception and illusion that result in our self-imposed blindness and the illusory knowledge that it postulates.

If we look closer, must we not realize that this state in which we can never possess any true knowledge—in which what we call our knowledge must inescapably be impregnated with falsehood—is itself the consequence of the fact that we identify ourselves with our ego, and allot to it a purely fictitious autonomy in which we attribute our thought, and our perception of ourselves and of the world, to ourselves, as though they derived from ourselves, and as though our thought was itself the determining factor of all things, including our own existence? In other words, do not our self-imposed blindness, and the illusory knowledge it postulates, stem precisely from that type of mentality which can in all sincerity make the kind of statements that Hamlet and Descartes make—statements behind which lurks the old Protagorean sophism about Homo mensura, man the measure of all things,[5] and which anticipate that triumph of the Demos expressed in such clichés as 'my view is as good as your view'? Because in the end, for this way of thinking, even the gods are nothing more than ideas in the human mind.

This, however, still leaves the more important question un-answered. For if how we perceive things with our untransmuted and unregenerate consciousness—and I will explain shortly what I mean by this—does not correspond to the reality or true nature of things, but simply reflects the self-deception and illusion that characterize our ego or our selfhood, what is the reality or true nature of things, and how is it that we fail to perceive it?

I should at this point make it clear that when I am speaking of things—of the true nature or reality of things—I am denoting things in the sensory world, visible things, what we call phenomena or appearances, or the world of nature. And when I speak of knowledge, I do not mean information about things; I mean understanding of what or who things are, of why they are, of their true identity and what they signify. And when I further say that our unregenerate consciousness, or what we might call our ego-consciousness, cannot perceive the reality or true nature of these things, I mean by this that the perception of our ego-consciousness is limited to the purely material and terrestrial aspects of these things—to their materiality, to those aspects which can be measured, quantified, reduced to what are thought to be mathematical equivalents or for which there is empirical evidence as these words are understood in the terminology of modern science. It is these aspects of things which our ego-consciousness regards as constituting their reality, and thus as capable of furnishing us with knowledge of the things themselves; whereas in actuality these exterior aspects of things do not constitute their reality and cannot furnish us with a knowledge of them.

To say this, however, is either to say quite simply that visible things possess no reality at all and are totally illusory; or it is to say that their reality is constituted by something quite other than those aspects of them which are accessible to observation when the agent of such observation is our ego-consciousness. But I have already said that things possess their own reality quite apart from whether we perceive it or not. Thus what I am affirming is, in fact, that their reality is constituted by something quite other than anything that can be perceived by our consciousness while it is still in an unregenerate, untransmuted state.

What, then, is this something? To start with, recognition of it demands that we read the book of nature, the Liber mundi, in a way totally different from that in which we have been taught to read it. It demands that we read it in a way similar to that in which the great spiritual expositors tell us that we should read the Bible or any other Holy Book, not according to its literal, outward meaning but according to its inner, spiritual meaning. This is to say that we have to learn to look on the world of natural forms as the apparent, exterior expression of a hidden, interior world, a spiritual world: all the phenomena of the world of nature represent or symbolize with things celestial and divine.

In this perspective, natural things are essentially effects, never causes, still less causes of causes. In no way in themselves can they suffice to account for their appearance or mutations, and in no way are they self-sufficient entities or themselves the cause of what they are. Each natural form derives from the cause which it manifests and represents, and is preceded and determined in every way by this cause. Each has its equivalent, or archetype, or Divine Name, on the spiritual plane, and is the external expression, the material extension of this archetype. In the whole visible, natural world there is nothing that does not express or represent something of a higher invisible world, the spiritual world. Without this rootedness in the spiritual world nothing could exist for an instant, for apart from the spiritual world nothing can have any existence at all. No visible thing—nothing belonging to the world of phenomena—possesses existence or being in its own right, and divorced from its inner and spiritual dimension and identity, it possesses no reality whatsoever, whether physical, material or substantive.

At this point I must insert a short parenthesis, to avoid misunderstanding. I said that natural events and phenomena are always effects, never causes, and that the cause of each such event or phenomenon, whether with reference to origin or to temporal permutations, is always spiritual, always supranatural. I emphasize this because our minds have become so dominated by linear thought and its mechanical ramifications that we tend to find it difficult to think in any other way. That is to say, we tend to envisage things, and causation and continuity themselves, in terms of an unbroken linear sequence according to which events and changes in the natural world happen because of other events and changes that have taken place in the past; and this notion of unbroken linear sequences and succession is used to explain the present state of things.

Yet this understanding of things in terms of a sequential cause-effect syndrome operating within linear time represents a total misconception of the structure of reality. In fact, it is not going too far to say that this linear model within which our thought has been conditioned to function is among the chief impediments, if not the chief impediment, to our understanding of anything that happens in the natural world, in the world of nature or of history. Causation and continuity are properties of the realm of archetypes or Divine Names. In the realm of events and phenomena there are connections, not causal relationships. All causality resides in the divine archetypes, in the incessant renewal of their epiphanies from instant to instant.

The recurrence of things in the world of events and phenomena consists in the recurrence of epiphanies. Thus the identity of a being, human or other, does not derive from any empirical continuity of its visible presence; it is wholly rooted in the epiphanic activity of its eternal archetype. In the realm of the manifest there is only a succession of likenesses from instant to instant. This of course implies a conception of the relationship between this world and 'the other world' quite different from that which we have become accustomed to. For this world *is* the other world. It is already the other world: the other world is perpetually engendered in this world, and from this world, which has no beginning and no end. To unlearn the concept of linear time and the notion of sequential cause-effect that lies at the root of linear thought itself, as well as all that they imply, is a *sine quâ non* of any genuinely scientific understanding of events and phenomena, natural or historical.

All that is in the natural world, then, from its minutest particle to the constellations, the whole and each particular of the animal, vegetable and mineral kingdoms, is nothing but a kind of representational theatre of the spiritual world, where each thing exists in its true beauty and reality. Each natural form is the centre of an influx coming from its divine archetype or theophanic Divine Name. Thus each natural form is the image—the icon or the epiphany—of its archetype, and by virtue of being such an icon each possesses an affinity with its archetype, it corresponds to it, symbolizes with it. And when I say it symbolizes with it I do not mean that there is any gap or disjunction between it and the archetype it symbolizes with. The one is the other, the archetype is the icon, the icon is the archetype, there is an indissoluble interpenetration of the one by the other. The numinous presence, of which the outward form of things is the image, is also present within it. Though there is a distinction, there is no dualism between the natural and the supranatural world. The spiritual world is not another world set apart from the natural world. It intermingles and co-exists with, and constitutes the invisible dimension of, the natural world. It is another world incorporated within the natural world. And this takes place, as Jan van

Ruysbroeck puts it, 'beyond time; that is, without before or after, in an Eternal Now . . . the home and beginning of all life and all becoming. And so all creatures are therein, beyond themselves, one Being and one Life . . . as in their eternal origin.'[6]

This brings us to the second question: if this is how things are, how is it that we fail so deplorably to see them as they are? There are two ways to approach the answer to this question—or, rather, two modes in which it may be answered. The first we might call historical, in that it consists in defining the emergence and nature of that premise which has bedeviled the thought and practice first of the modern western world and now of virtually the whole world. The second we might call transhistorical, in that it consists in an archetypal drama played out on the gnostic and mythic plane of the human spirit.

To speak briefly of the first. It is connected with certain intellectual developments within the European Christian world which have to do with changes in the relationship between what one might call metaphysical knowledge—knowledge of the supranatural and uncreated world—and physical knowledge—knowledge of the natural and created world—that took place within the Christian theological consciousness in the later mediaeval period, particularly in the thought of the Scholastic theologians and philosophers. Or, rather, it might be more accurate to say that they received explicit formal expression in this period, although they had been incubating in the Christian consciousness for some centuries prior to this, in both the Greek East and the Latin West.

In very general terms, one might say that these changes represent the displacement of a unitary approach to knowledge by a bifurcated, dualistic approach. In the unitary approach there is no division or separation between knowledge of the supranatural and the uncreated world on the one hand and knowledge of the natural and created world on the other: the two run in tandem, are harnessed together; they constitute a single form of knowledge, a single science. This is because the natural and created world is perceived as the embodiment, the material and visible prolongation of realities that are immaterial, spiritual and uncreated, so that there is no way in which we can understand or possess a true knowledge of the natural and physical world without a prior understanding and a prior knowledge of the supranatural and divine world, for the simple reason that, as I said, divorced from its inner and spiritual dimension and identity no visible thing, nothing belonging to the world of phenomena, can possess any reality at all.

We can attain such an understanding and knowledge of the supranatural and divine world both indirectly and directly. We can attain it indirectly from the Holy Book—the Liber revelatus, the Book 'descended from heaven'. The truths of Revelation, although given form in a particular historical and cultural context—namely, where the Christian tradition is concerned, in the life of Christ as this is shown forth in the Gospels—nevertheless correspond to eternally present divine realities; they are a revelation of the true nature of things, of what is entirely normal, not exceptional. Certainly, these truths have to be unveiled from the literal sense in which they are, as it were, concealed in the Holy Book itself. The objective data are provided by the revealed Holy Book, the revealed and revealing divine Logos; but the question is to know their true meaning, their spiritual meaning, not simply their literal meaning. 'The letter kills, but the spirit endows with life.'⁷

This is not to invoke the mediaeval theory of the four senses of Scripture, the literal, moral, allegoric and anagogic. It is, though, to presuppose what one might call a theosophia: the gnostic or visionary perception of a whole hierarchy of spiritual universes—universes which are not to be disclosed by means of syllogisms, since their unveiling requires a certain mode of knowing, a hierognosis, which combines the reflected knowledge of the data given by Revelation and the most personal inner experience; for without such experience, all that can be conveyed is a mere collection of concepts and abstract formulas, more or less arbitrary and essentially fugitive. One might say that the divine revelation is the light that makes it possible to see, while the inner experiential vision of the gnostic is the light that sees. To ignore the first—the divine revelation—is to remain permanently in the dark. Not to attain the second—the inner experiential vision—is to remain blind.

Here I would like to forestall a possible objection. I said that truly to read the book of nature, we have to read it in the same way as we read the Holy Book, the *Liber revelatus*. This, of course, is to assume that both are read in the light of the *theosophia* of which I am speaking, so that the interpretations which are given of either have a transtemporal or transhistorical validity. This assumption and its rider will be of a kind that the modern scientist cannot accept. But they are also rejected by an increasing number of theologians who, persuaded perhaps by the awareness that all hypotheses postulated by modern scientists as a result of their reading of the book of nature cannot at best be other than purely provisional and non-definitive, are led to apply the same conclusion to the reading of the Holy Book and to assert that all interpretation of revealed scripture must likewise be purely provisional and non-definitive.

They would support this conclusion by saying that just as the way in which the modern scientist reads the book of nature is determined by subjective temporal, contextual, and other parameters, so is the way in which the theologian reads the Holy Book; thus the interpretations of the latter are just as dependent on these parameters as the hypotheses of the former, and both will have a certain working validity only in and for their time. This way of looking at things is reinforced in both cases by the common acceptance of the concept of linear time of which I have already spoken, allied to a corresponding notion that there is a certain evolution of consciousness that tends to be part and parcel of it, making it permissible to use such phrases as 'the dawn of human consciousness', 'the emerging consciousness of our times', and so on. Man is seen as essentially a temporal being, and his thought as necessarily determined by his place in history, in such a way that it can, with the passage of time, become outmoded or obsolete, in need of replacement by more contemporary types of thought.

This attitude effectively undermines, if it does not negate, the understanding that, just as revelation itself, so its spiritual interpretation or hermeneutic—what we call doctrine—has its origin not in the historical order but in the transhistorical order; that it does not pertain to the domain to which scholars can apply their critical historical criteria; and that it is no more a merely sociological or cultural phenomenon than is a human being. Certainly, such interpretation and its articulation in doctrinal forms do reflect, and so indeed are limited by, the temporal consciousness of those responsible for them—this is one of the reasons why they must always be approached in an apophatic manner. But the divine gnosis of which they are the interpretation and articulation transcends these limitations and is eternal, being as it is the effulgence of divine light and life.

Again, with our current sense of history, as of time, we tend to see the irruption of such gnosis into this world as an event taking place at a certain moment 'in the past', once and for all and irreversible. But when truly understood, this gnosis and the anterior realities of which it is the mirror are not, and can never be, 'of the past'; they are always 'in the present' (*instantem*). Correspondingly, spiritual interpretation of this gnosis can also never be of the past, but must be 'in the present', since it transcends time by re-attaching all temporal manifestation to its non-temporal source. Thus its articulation has, like the incarnation of the Logos in the historical Jesus, a transhistorical dimension, and for those who have eyes to read it, it can never become obsolete or lose its symbolic function as the authentic expression of divine gnosis.

Admittedly—and here I return to the main theme—the *theosophia* of which such articulation is the expression presupposes both the reality, and the possibility of direct personal perception, of a more-than-human body of knowledge—what St. Augustine calls 'Wisdom uncreated, the same now as it ever was, and the same to be for evermore'[8]—which pre-exists all interpretation and articulation. And this in its turn is to presuppose that inherent in each human being is an organ of vision, of intellective or imaginative intuition, which when activated is capable of perceiving and experiencing the realities of the supranatural and divine world.

This organ cannot, of course, be the reason. The reason itself cannot have a direct vision or experience of anything. It can operate only from a given starting point, or from given starting points, which we have to assume as a condition of being able to set the progress of reasoning in motion. So what is implied is that we possess within ourselves an intellective and visionary organ that is superior to the reason, and that it is this organ, or power, that is capable of perceiving the inner and spiritual reality of things.

Such an organ—it may be called the spiritual or angelic intellect—although present within us is initially, and sometimes chronically, present in a latent or potential or passive state—why this is so will also become clear later—so that to all intents and purposes it is not operative within our consciousness until it has been brought from a state of passivity into a state of activity, from a state of potentiality to a state of being fully operative. This is why I spoke earlier of our ordinary or ego-consciousness as being 'unregenerate' or 'untransmuted'; for what regenerates and transmutes our consciousness is precisely this process of bringing our angelic or spiritual intellect from a state of potentiality to a state of active realization, so that it becomes not merely operative in our consciousness but the determinative and transforming agent of our consciousness, conferring on it the capacity to perceive in things those inner and spiritual qualities to which our unregenerate and untransmuted consciousness is totally blind.

At the same time, this regeneration and transmutation of our consciousness has as a direct consequence the regeneration and transmutation of our sense-organs as well, so that they, too, are changed from being what Herakleitos calls 'false witnesses', incapable of registering the spiritual and numinous qualities of things, and become capable of participating in the spiritual vision of our reborn consciousness and of sharing in its now undistorted perception. For as the veils are lifted from our consciousness, so the veils are lifted from our sense-organs; just as, correspondingly, when our consciousness is closed to divine life and light, our senses are impervious to them as well.

This unveiling of our consciousness so that it ceases to be impervious to divine life and light—its transmutation and regeneration—is, as I said, a process that goes hand in hand with the realization or actuation of the potentiality of our spiritual intellect. But this process is far from being an automatic one. On the contrary it can be accomplished only on condition that we pursue a long, strenuous and often extremely taxing course of spiritual practice and purification, inner and outer, mental and physical. Christianity, like every authentic sacred tradition, also possesses its own initiatory and mystagogical discipline, gnostic and ritual; and it is through participation in such a discipline, and only through participation in it (except in the few cases that constitute the exceptions which prove the rule), that we can begin to penetrate into the hierarchy of spiritual universes of which previously we had been totally ignorant or had only accepted 'on faith', as a kind of theoretical basis for our deliberations. And the further we penetrate into these realms, and the more our consciousness is opened to the influx of divine life and light, the more we can decipher the spiritual meaning of the data given in the revealed Holy Book, the Liber revelatus; and the more we can do this the more we are able to attain a direct perception and knowledge of the realities of the supranatural and divine world.

This means that simultaneously and correspondingly we are also initiated into a true reading of the book of nature, the Liber mundi, for these realities constitute the immaterial, spiritual and uncreated realities of the forms of the natural and physical world; they embrace the archetypes of which these forms are the apparent, exterior expression. And because when our consciousness is spiritualized by the reawakening of our supreme cognitive faculty—the spiritual intellect—our senses are also transmuted and spiritualized, this in its turn means that we are able to perceive through our physical eyes the symbolic function that natural things possess by virtue of their correspondence and interpenetration with spiritual things. We are able to perceive their inner and spiritual dimension and identity.

It is precisely our capacity to perceive this symbolic function of natural things—to perceive the numinous presence of which each natural form is the icon—that is increasingly eclipsed by those intellectual developments that took place in the Christian, and hence by and large European, consciousness in the later mediaeval period. These are typified by the invention of a particular concept, the concept of the 'double truth' promoted in the thought of Scholastic and subsequent theologians and philosophers. What exactly is this concept?

We have seen that in the unitary approach to knowledge, there is no separation or division between knowledge of the supranatural and uncreated world and knowledge of the natural and created world; no separation between the truth revealed in the Holy Book and the truth revealed in the book of nature: the illumined gnostic, his consciousness transmuted and regenerated, can perceive the same supranatural and divine realities expressed equally in and through both of them, and his knowledge of both of them derives from, and is dependent upon, his perception of the divine realities which they both enshrine.

As a consequence of the intellectual developments to which I have referred, this understanding of things is displaced and rendered existentially ineffective, with the result that unitary knowledge, and the vision of the natural world that is part and parcel of it, are also displaced; for the type of consciousness which they demand is eclipsed, and is replaced by another type of consciousness. For, first, the understanding of man as a triune being of spirit, soul and body gives way to the notion that he is but a dyadic being of soul and body alone. This in its turn means that the faculty which according to the unitary theory of knowledge constitutes our supreme cognitive faculty—our angelic or spiritual intellect—is no longer recognized as a power inherent in human nature, a power altogether superior to and independent of the reason, since it possesses an intrinsic spiritual potentiality. Now the intellect (the word is still used) is seen as no more than a higher aspect of the reason itself. Essentially intellect and reason now describe one and the same power, there is no spiritual visionary organ in man distinct from his reason, and the mode of knowledge proper to man is reasoning or discursive knowledge, a mode within everyone's scope without the operation of any initiatory grace, for by definition everyone possesses a soul and by definition that soul is a rational soul.

It follows from this that the visionary perception of divine and supranatural realities which is a presupposition of any genuine knowledge of the natural world—for without it we are blind to the inner and spiritual dimension of things—is now regarded as being beyond the reach of the human intelligence. It is true that it is still assumed that natural forms have an analogical resemblance to supranatural forms; but this is not in the least taken to mean that we cannot obtain a knowledge of natural forms simply by studying them in themselves, without any reference to supranatural forms. On the contrary, it is now thought that this is the only way in which we can study them, since any direct knowledge of their supranatural dimension is assumed to be beyond our scope. What is within our scope is said to be limited to a knowledge of natural things as they are perceived in the natural light of the human reason.

What begins as the scholastic assertion—already in itself representing a radical inversion of the norms of knowledge—that the world we perceive through the senses is the primary source of human knowledge (quod in intellectu est, primo in sensu erat), is now elevated into the dogma that this world is the only source of human knowledge, a knowledge, moreover, that is thought to be entirely valid. It is the growing ascendancy of this misconceived dogma, first in the European and subsequently in the extra-European mind, that has ensured the progressive materialization of every aspect of our culture.

What, though, in the meantime has happened to that other source of truth, the Holy Book, the Liber revelatus? Does the fact that it is now thought that the human reason can obtain a perfectly valid knowledge of the natural world, by abstracting it from the observation of those aspects of this world which we can perceive by means of the senses, imply that the truths of divine Revelation are regarded as superfluous? This is not the case at all. But what has happened is that the whole attitude towards, and understanding of, these truths have undergone a similar change. For these truths, although still regarded as valid (since God has revealed them to us), are now said to be beyond human capacity to know in a direct, experiential manner, through noetic penetration into the spiritual universes of which the literal form of the Holy Book constitutes the outward expression. And they are regarded as beyond human capacity to know, because human beings are no longer thought to possess an organ of vision through whose actuation they can be known. The things of faith—the truths of Revelation—which must be believed by all, are equally unknown by all, and there can be no direct experiential knowledge of them.

This might not have presented a problem had the truths of revelation always agreed with the conclusions which the human reason, now granted a charter to operate quite independently of the truths of revelation, derived by a process of abstraction from the observation of natural phenomena. Unfortunately, however, this was by no means always the case. Hence some way had to be found of accommodating both the conclusions of the reason and the truths of revelation, even when they appeared to conflict with each other; for now both theologians and philosophers were committed to accepting that both sets of truths, even when they appeared to conflict, could be valid.

The only way that could be found to effect the accommodation in question was to divide the sphere of revelation from that of reason, to divide faith from philosophy and science, metaphysics from physics. St Thomas Aquinas, following on the Jewish philosopher Maimonides and others such as Alexander of Hales, Bonaventure and Albert the Great, clearly asserts this distinction: on the one hand there is faith, which is assent to something because it is revealed by God; and on the other hand there is science, which is assent to something because it is perceived to be true in the natural light of human reason. The two departments are separate, the truths of the one being valid in one sphere, the truths of the other in another sphere.

It is this concept of the double truth—this duplicity in the true sense of the word, or what we might call 'double-think'—that constitutes the major premise that has to be reversed if ever we are to escape from the clutches of our materialist world; for it is this concept that constitutes the bedrock of the major thought-paradigms responsible for shaping the course and character of this world, and that is inextricably built into them. It is the initial bifurcation, or splitting asunder, that has given rise to the whole crisis of fragmentation which now threatens to disrupt whatever is left of anything that can be called civilization. It marks the decisive breakthrough of that type of consciousness I have called our ego-consciousness—the consciousness which, being closed to divine light and life, simply reflects the self-deception and illusion that characterize our uprooted selfhood; and simultaneously it marks the opening of the door to the progressive secularization and profanation of virtually every aspect of our life, public and private, philosophical, scientific, political, social, educational and even domestic.

For when, in obedience to the dictates of a theology that has separated the order of supranatural knowledge from the order of natural knowledge, philosophy and hence science declare their independence of the truths of Revelation—their independence of the Holy Book and the spiritual hermeneutic that unveils its meaning; and when they further declare that the human reason is capable in its own right of acquiring a valid knowledge of things, there can be no preventing the breakthrough of this type of consciousness or the consequences that follow inescapably from its ascendancy.

That the ascendancy of the ego-consciousness is in fact written into the status now accorded to the human reason, and the license assigned to it, is due to the very nature and function of the reason itself. We have already noted that the reason itself cannot have a direct knowledge of anything. It can operate only from given starting points, and it reaches its conclusions by deducing them logically from these starting points. If these starting points are provided by the understanding of reality that accords with the type of perception intrinsic to our spiritual consciousness, the conclusions that the reason will reach with regard to the forms of the natural world will be of one kind; if these starting points are provided by the understanding of reality that accords with the type of perception intrinsic to our ego-consciousness, its conclusions will be of a totally different order. This is to say that a science of the natural world can be termed rational only in so far as its conclusions follow logically from the premises that constitute the starting points from which the reason operates, irrespective of the type of consciousness that determines these premises and therefore its whole subsequent structure, theoretical and methodological. There is no science that is rational as such, or that is more rational if it takes as its starting points one set of premises rather than another set. Nor can the human reason in itself establish these premises, or their rationality, or their truth or falsehood, nor can it acquire a knowledge of the natural world. There is no way in which the natural light of human reason can in itself attain a knowledge of anything. This is why the concept of the double order of truth, promoted by the Scholastic and other philosophers and subsequently built into virtually every aspect of our intellectual and other life, represents such a distortion of things.

This being the case, what precisely is involved in promoting the idea that the human reason is perfectly competent to investigate things without reference to any Holy Book or to its corresponding spiritual hermeneutic, and that it possesses in its own right the capacity to obtain a valid knowledge of the physical world? What is involved in claiming that it is perfectly legitimate to separate the sphere of reason from that of revelation, physics from metaphysics, science from faith? As I have already indicated, what is involved is a radical inversion, not to say perversion, of the norms of knowledge. For now it is no longer recognized that in order to formulate a genuine knowledge of things the reason has to accept as its starting points propositions that accord with the perception of our spiritual consciousness. On the contrary, it is now asserted that the reason can formulate an equally valid knowledge of things by taking as its starting points propositions that accord with the perception of our ego-consciousness, a perception which simply reflects the limitations and characteristics of our unhallowed, self-enclosed selfhood. This means that the premises now effective for philosophy and hence for science are no longer those of the *hierognosis* or *theosophia* of unitary knowledge; they are merely those which the human mind, cut off from a direct perception of spiritual realities, happens to invent or adopt in a purely arbitrary and subjective manner. Given that such a mind is by definition in thrall to self-deception and illusion, and is exposed to every kind of distortion, the conclusions of the reason that operates in accordance with the premises it invents or adopts will be conditioned correspondingly.

In other words, to the purely subjective and unsanctified human consciousness is attributed an autonomy that establishes it as the determinative factor not only of the norms of knowledge, but even of the norms of human existence itself: we are back where we started, with that type of mentality which can in all sincerity make the kind of statements that Hamlet and Descartes make, or that Protagoras makes, and with the agnosticism and materialism which are part and parcel of it. That an agnostic and materialistic science of nature is a contradiction in terms, and that its findings will necessarily correspond to the living reality of nature as little as a corpse corresponds to the living reality of a human being, will be clear from all that I have already said; just as it will also be clear that our tragedy, and the tragedy of the world we live in, are due to nothing more, or to nothing less, than that we prefer the desolation of our own destruction to having to expose ourselves to the trials, dedication and love that are the touchstones of our regeneration.

For it is because we have chosen and continue to choose to live according to this delusion of the double truth that we have ended up by shattering our ancestral universe, our spiritual cosmos, into a thousand fragments and have ejected ourselves into a world that is as nightmarish as it is artificial. It is this duplicity or double-think that lies at the root of what has now become our endemic state of schizophrenia. It is this that permits us to say that we are Christians, or Moslems, or Buddhists, or whatever, and yet to live according to values and standards and ideas that not only have nothing to do with any religion but are entirely contrary to every form of spiritual life and practice. It is this, finally, that allows us not simply to tolerate, but actively to promote, a type of science that inevitably desecrates every area of life on which it impinges because desecration is written into the very view of nature according to which it operates; for this view is itself the progeny of this same misbegotten concept of the double truth which would have us believe that nature is a self-subsistent reality, independent of God, with nothing holy or sacred about it, and that it is quite possible to acquire a valid form of knowledge by investigating it as such.

So long as we fail to realize that this view of nature is a complete misconception, and that what we call our empirical and experimental knowledge of facts is itself an essential part of our ignorance, we will continue to desecrate the earth and everything on it without even the slightest awareness of what we are doing or why we are doing it. And we will go on failing to realize that this view of nature is a complete misconception for so long as we also fail to realize that the words of Christ, 'In so far as you did it to one of the least of these my kindred, you did it to me',⁹ apply not simply to his human kindred, but to every natural form of life and being. For every natural form of life and being, down to the most humble, is the life and being of God.

Moreover, the fact that we have been increasingly persuaded over the last centuries, by our submission to the norms and propaganda of the modern scientific mentality, to regard as knowledge only that which has direct reference to those aspects of the natural world that are accessible to the senses has meant that we have become progressively dominated by an insatiable craving to experiment in ever more extreme and demented forms of sensation. For this scientific mentality and its propaganda have not only stimulated our purely profane curiosity beyond all measure, human or divine, but at the same time they have compelled us to accept as axiomatic that such curiosity can be satisfied only by an experimentalism which involves testing hypotheses against what is called empirical evidence, the evidence of the senses.

This cult of experimentation pervades virtually the whole scholarly world, in one form or another. Yet it is a singularly inept cult. For in so far as it is practised as a means of assessing an hypothesis by testing it against empirical evidence, it simply engages its votaries in the specious rite of arguing in a circle. This, as I have pointed out, is because what a scientist, as anybody else, will think constitutes empirical evidence is already determined for him by his prior commitment to the presuppositions built into the conceptual paradigm that happens to dominate his consciousness. But it is precisely these presuppositions, and his commitment to them, that also determine the formulation of the hypothesis which he now wishes to assess by checking it against this evidence. Yet because of the prestige accorded to this cult, the addiction which it represents is more sinister than most of the other forms of addiction by which our contemporary society is afflicted, for it lies at the root of virtually all of them.

In addition, it has to be remembered that the kind of specialization that characterizes virtually every form of mental activity deployed in the modern world, and especially every form of mental activity deployed in the world of modern science, requires the exercise of only a fraction of our intelligence; the greater part of our intelligence, never called into operation, simply deteriorates or atrophies. As a result we are by and large incapable of entering into realms of thought which lie beyond the extremely narrow scope to which our intellectual understanding, and hence our lives, are now confined, so we cannot but react negatively when we are invited or challenged to enter into those realms, and may even pretend that such realms are non-existent or belong to the world of fantasy. In one sense this is our major problem—how to reactuate powers of our intelligence that now lie dormant, so that we can once more become aware of those realms of thought, and of the realities they mirror, which are now excluded from our field of vision. All that can be said here is that the first step towards such a reactuation would have to be the unlearning of much of what is now called knowledge, and the freeing of ourselves from the presuppositions on which it is based.

No one—or at least hardly anyone—wants to know of such things, and one is regarded as slightly barmy or hysterical when one points them out, and one will get little thanks for so doing. Yet whether we recognize it or not, the truth is that we are involved in a drama that is first of all a spiritual drama, an eclipsing of our spiritual vision; and it is this self-imposed blindness that has resulted in the whole process of disintegration of which the intellectual developments of which I have just been speaking, as well as their consequences in every sphere of our life, are the outward manifestations and symptoms. To conclude, I will say something about the nature of this drama—something that will represent in gnostic and mythic terms a recapitulation of what I have already said.

At the beginning of this paper, I remarked that the reversing of the premises of the type of thought lying behind our present plight entails no less than a reversing of a process of ignorance; and that this reversing of a process of ignorance is in its turn the prelude to a regeneration which itself is a return to a state of being and consciousness which can only be described as paradisal. What is this paradisal state, and what does it signify to be exiled from it?

In its simplest terms, to be in a state of paradise is to be free from bondage to self-deception and illusion, and to perceive things, ourselves included, as they truly are, and not as they appear to be through the distorting mirror of our unregenerate minds. This, as I have already explained, means that our perception of sensory things—of created things—will simultaneously embrace the perception of their divine and celestial reality, so that what we see with our eyes when we are in this state are celestial beings, living manifestations of divine life.

At the same time, in the paradisal state, although we possess this spiritual perception of things, we are aware that the Subject active in all our acts of knowing and perceiving is not our own selfhood, but is the Divinity Itself, is God Himself. Or rather, we might say that in this state our personal selfhood is conscious of itself as inseparable from the divine Principle. Our selfhood is transparent to this

Principle: its life and light are the life and light of God Himself.

Paradise—the garden—is, then, divine inspiration flowing into us directly and without intermediary. It is knowledge in the true sense, pure spiritual science. This means that when we are in this state of being, we are not simply in paradise: we *are* paradise. We are the state in which we are; our state of being—our *modus essendi*—corresponds to our state of knowing—our *modus intelligendi*. Paradise is, then, our inner state; and what this inner state—our mode of being—allows us to experience is an infinity of perceptions derived from the divine Principle active in us. That is why even if we are in paradise, and even if we are paradise, paradise is not ours, for all our perceptions are activated in us by God.

The loss of this inner state—of this paradisal state—is described in Christian terms as the 'fall', an event not of the past but one in which we are involved at every moment of our lives. To see what is meant by this, we have to remember that there is a vital distinction between our true being and what we become when we identify ourselves with our own selfhood. Our natural state is our paradisal state, with the consciousness and knowledge that go with it; and we saw that in this state we are conscious that our personal selfhood is rooted in its divine Principle, and that this Principle is the subject of all our acts of knowing and perceiving. But we also possess the possibility of taking another path and of regarding ourselves as existing independently of God: we can assert the autonomy of our selfhood and identify ourselves with that. We can think that we are ourselves of and through ourselves, as self-subsisting beings. We can think that we are the agents of our own lives and of our acts of knowledge and perception. We can become the victims of the terrible illusion that one's own self is sufficient to itself in order to be itself. Swollen like Faustus 'with cunning, of a self-conceit',[10] we, like him, can claim that our souls are our own.

Yet on the day that we do this—on the day on which we assume that we can perceive and know things by ourselves and can investigate the secrets of our own being and of the natural world by ourselves—we die to our celestial nature and to the knowledge and consciousness that go with it. For once we attribute our perception of ourselves and of nature to ourselves, as though it derived from ourselves, we will automatically and inevitably close our consciousness to the influx of divine light and life. God will cease to be the active Subject of our acts of knowing and being, our celestial perception of things will be destroyed, and we will substitute for it, as the purveyor of what we now call knowledge, a perception that can see in things nothing more than their purely psychic and material aspects.

When we become attached to our selfhood as a dissociated, independent reality, two things happen as a consequence. The first is that we cease to realize the difference between our natural state—our paradisal state—and the fictitious autonomous state which we now attribute to ourselves and with which we now identify ourselves. One could put this in another way and say that we cease to identify ourselves with our real being and identify ourselves instead with that which we think we are. That which I truly am—my real identity—is always full and complete and indestructible. That which I think I am when attached to my selfhood is nothing more than an invention of my own mind, an illusion to which I attribute reality and which I regard as a concrete entity. It is nothing more than a certain bundle of mental and physical accidents, thoughts and feelings and sensations, derived from heredity, education, environment, and a thousand other transient influences and activities, temporarily brought together but changing in such a manner that there is never a moment with reference to which I can say of myself that it is myself: for as soon as I ask what it is, it has become something else. There is in fact on this level no true entity I can call myself; and my ego-consciousness, according to which I regard my ego as an entity, is itself as fictitious as the ego to which it attributes reality, since it is merely a sequence of reactions in whose current we are immersed and by which we are endlessly swept along.

Thus when we assert our independence of God, what we are really doing is centring our attention on a purely fictitious self and attributing to it a reality, and a permanence, which it does not and cannot possess. And it is this false self, or ego, the constantly renewed creation of our ignorance, that becomes the focus of all our illusions, our passions and, it might be added, our sins. It is the independence and autonomy we ascribe to our selfhood that is the crux of our aberration; for this selfhood, being itself a deception, is also the agent of every kind of trickery and falsehood. And it is our attachment to it that precipitates our fall.

The second thing that happens as a consequence of such an attachment to our selfhood corresponds to the first: we cease to be able to perceive the true nature of things, their spiritual and celestial reality, and we identify them instead with what we think they are. But to identify them with what we think they are is to project onto them the same kind of fictitious and illusory identity as that with which we have already identified ourselves. In the end, we cease wanting to know of the existence of what is spiritual and celestial, whether in ourselves or in anything else, and we even deny that we or anything else can possess such qualities. Indeed, we reach the point of thinking that the world of three dimensions which we perceive with our ego-consciousness and with the aid of our five senses is the natural world and that it possesses its own self-subsistent existence.

Our fall, then, is our sundering of our links with the divine and an increasing attachment to the norms of our non-spiritual, non-spiritualized selfhood. It is a limiting of ourselves to our selfhood, a falling in love with ourselves, our surrender to an auto-erotic concupiscence, to an obsessive narcissistic psychosis. We commit, in short, a kind of suicide. And as a consequence, we are expelled from the garden—the paradise—that was, and is, ourselves, in our true being, in our state of transparence to the divine. And since we were—since we are—this garden, our expulsion from it, self-induced, is nothing less than expulsion from ourselves, alienation from our inner being, the despiritualization of human existence, a lapse into a sub-human state. And this expulsion is in its turn an entry into a state of profound illusion. For although the dissociated, fictitiously autonomous selfhood with which we now identify ourselves appears to us to be real, it is, as I said, essentially unreal and illusory, so that our subjection to it inevitably leads, as I have also said, to our enslavement to an illusory world entirely of our own invention. And there can be no more abject and degrading situation in which to find ourselves than that of being enslaved to our own inventions. To be in such a situation is in truth to be in hell, for in that situation we have lost our reality, and that precisely is hell.

In the light of what I have been saying, we can see why it is that the fall may best be understood not as a moral deviation or as a descent into a carnal state, but as a drama of knowledge, as a dislocation and degradation of our consciousness, a lapse of our perceptive and cognitive powers—a lapse which cuts us off from the presence and awareness of other superior worlds and imprisons us in the fatality of our solitary existence in this world. It is to forget the symbolic function of every form, and to see in things not their dual, symbiotic reality but simply their non-spiritual dimension, their psycho-physical or material appearance.

Seen in this perspective, our crime, like that of Adam, is equivalent to losing this sense of symbols; for to lose the sense of symbols is to be put in the presence of our own darkness, of our own ignorance. This is the exile from paradise, the condition of our fallen humanity; and it is the consequence of our ambition to establish our presence exclusively in this terrestrial world, and to assert that our presence in this world, and exclusively in this world, accords with our real nature as human beings. In fact, we have reached the point not only of thinking that the world which we perceive with our ego-consciousness is the natural world, but also of thinking that our fallen, subhuman state is the natural human state, the state that accords with our nature as human beings. And we talk of acquiring knowledge of the natural world, when we do not even know what goes on in the mind of an acorn.

This dislocation of our consciousness which defines the fall is perhaps most clearly evident in the divorce we make between the spiritual and the material, the esoteric and the exoteric, the uncreated and the created, and in our assumption that we can know the one without knowing the other. This is to say that if we acknowledge the spiritual realm at all, we tend to regard it as something quite other than the material realm and to deny that the Divine is inalienably present in natural forms or can be known except through direct perception which bypasses the natural world, as though the existence of this world is, spiritually speaking, negative and of no consequence where our salvation is concerned.

This other-worldly type of esotericism only too often degenerates into a kind of spiritual debauchery, in the sense that it has its counterpart in the idea that it is possible to cultivate the inner spiritual life, and to engage in meditation, invocation and other ritual practices, whether consecrated or counterfeit, while our outward life, professional or private, is lived in obedience to mental and physical standards and habits that not only have nothing spiritual about them but are completely out of harmony with the essential rhythms of being, divine, human and natural. We should never forget that an authentic spiritual life can be lived only on condition, first, that the way in which we represent to ourselves the physical universe, as well as our own place in it, accords with the harmony instilled into its whole structure through the divine *fiat* which brings it into, and sustains it in, existence; and, second, that in so far as is humanly possible we conform every aspect of our life, mental, emotional and physical, to this harmony, disengaging ourselves therefore from all activity and practice which patently clash with it. If we offend against the essential rhythms of being, then our aspirations to tap the wellspring of our spiritual life are condemned to fruitlessness, or, in some cases, may even lead to a state of psychic disequilibrium that can, in truth, be described as demonic.

Correspondingly, the divorce between the spiritual and the material means that material forms are regarded as totally non-spiritual, and thus either as illusion or as only to be known through identifying their reality with their purely material aspects. Such a debasement of the physical dimension of things is tantamount not only to denying the spiritual reality of our own created existence, but also, through depriving natural things of their theophanic function, to treating a divine revelation as a dead and soulless body. And in this case it is not only of a kind of suicide that we are speaking; we are also speaking of a kind of murder.

It is just as dangerous to think we can attain a knowledge of God while ignoring, or even denying, His presence in existing things and in their corresponding symbolic rituals as it is for us to think that we can attain a knowledge of existing things while ignoring, or even denying, the divine presence that informs them and gives them their reality. In effect, there cannot be a knowledge of the outward appearance of things—of what we call phenomena—without a knowledge of their inner reality; just as there cannot be a knowledge of this inner reality which does not include a knowledge of the outer appearance. It is the same as with the Holy Book: the integrality of the revelation cannot be understood simply from its letter, from its outward literal sense; it can be understood only when interpreted by the spiritual science of its inner meaning. At the same time this inner meaning cannot be perceived except by means of the letter, of the outward literal sense. There is an unbreakable union between the esoteric and the exoteric, the feminine and the masculine, between the inner reality of a thing and its external appearance. And any genuine knowledge of either depends upon both being regarded as integers of a single unified science.

Such knowledge is not therefore independent of the exoteric aspect of things. It is not a superior, esoteric doctrine which does not need to take into account the outer appearance. It is knowledge of the invisible dimension of things through which alone their outward appearance can be rightly understood. To attempt to attain a knowledge of either inner or outer as if the one were or could be independent of the other is to condemn oneself to an ignorance equivalent to that of Adam and Eve when, exiled from paradise, they knew themselves to be stripped bare of their natural vestments of intelligence and wisdom. Not to understand this is again to succumb to the lie of the double truth—to the lie that has vitiated our whole culture and has brought us to the edge of nemesis. And unless we overcome this lie, all our other efforts to escape that nemesis will be in vain.

This means that we have to disabuse ourselves of the idea that what we have been schooled over the last centuries to regard as knowledge does in fact constitute knowledge, as well as of the idea that knowledge can be obtained in the ways in which we assume we can obtain it. True knowledge can in no way be the acquisition or discovery of an individual, or of a group of individuals. It cannot be found through experiment (which always means treating things with violence), or through research (which always means interfering with nature), or by following the analytical path, for it has nothing to do with anything that can be chopped into pieces, split into sub-divisions, dissected, or broken down into constituent parts, or that assumes that things have to be compartmentalized or fragmented in order for something to be known about them. It cannot be investigated, or pinned down, or classified, or formulated into laws as Newton formulated the law of gravity, as if the earth was something to which we are bound because an impersonal objective force compels us to be so. Nor can we possess true knowledge unless we have personal experience of it; and it totally eludes every form of specialization, as it does any attempt to infer it from the observation of many particular instances.

For, as I said at the beginning of this paper, the so-called knowledge we think we can obtain by such means is the result of our attempt to know what we do not know and what we think is not known. It is a product of our not-knowing, of our ignorance, and our ignorance is inextricably built into it. And this ignorance that is built into all our so-called knowledge, and which prevents us from seeing things as they really are and from perceiving their true nature and identity, ultimately has its roots in our ignorance of who we ourselves are, and of what constitutes our true nature and identity; for, clearly, unless we first know who we ourselves are, we cannot know what anything else is either. And knowledge into which such ignorance is inextricably built cannot by definition constitute true knowledge. It must inevitably be tainted with this ignorance and hence with falsity. And knowledge tainted with falsity cannot constitute true knowledge. In other words, the human mind, without enlightenment from a more-than-human source, cannot attain a valid form of knowledge.

This in its turn means that if we are to see things as they are, we have to free ourselves completely from this kind of pseudo-knowledge and from the methodologies that go with it. We have to free ourselves from all that we think we know, empty our minds of all that we think we know, of all the conceptions we have formed as a result of going in pursuit of a knowledge we think we can obtain by any of the means of which I have just spoken. We have to become ignorant of all knowledge we think we have obtained through our own efforts. For true knowledge cannot be acquired by any of these means, and still less can it be confirmed or verified by any such means. Anything we can discover for ourselves by such means is pseudo-knowledge, non-wisdom. True knowledge has its source in the wisdom that is the lifeblood of all things and where everything is already known. It is not therefore something that is not known. It is not even something that we do not know. We do know it—it is our lifeblood—only we have forgotten it and lost it, just as we have forgotten and lost our own reality. If we can recover our own reality we will also recover this knowledge, for the two go hand in hand, this knowledge is part and parcel of who we are, in our true being. If we recollect who we are, we will also recollect this knowledge.

True knowledge, then, is something that is given to us, but we can perceive it only when we are in a condition to perceive it. It is not in any way subject to our will, we cannot capture it, we can only be suffused with it, embraced by it, immersed in it. It is a light that dispels the darkness of our ignorance, but a light that remains invisible to those who have eyes but cannot see. And we will never have eyes that can see while we are still dazzled by the glare of pseudo-knowledge. We cannot go on chasing the sirens and will-o'-the-wisps of pseudo-knowledge, let alone conforming our everyday life to its fallacies, and at the same time expect holy Wisdom to make us her tabernacle. For this to happen, we have to attain a new state, a state of unknowing which, contrary to the negative not-knowing, frees us from bondage to our ego-consciousness and to its stream of hallucinatory and dismembering thought, and allows us to perceive the seamless robe of nature in all its pristine integrity. Only then, through this act of self-recollection—which is also an act of remembering—will Wisdom reveal herself, will she unveil her presence in every natural form of life and being. Only then will we begin to see the beauty at the heart of things. And this is the road to the recovery of paradise. This is itself the overture to paradise.

WENDELL BERRY

Going to Work

❧

I.   To live, we must go to work.

II.   To work, we must work in a place.

III.   Work affects everything in the place where it is done: the nature of the place itself and what is naturally there, the local eco-system and watershed, the local landscape and its productivity, the local human neighborhood, the local memory.

IV.   Much modern work is done in academic or industrial or elec-tronic enclosures. The work is thus enclosed in order to achieve a space of separation between the workers and the effects of their work. The enclosure permits the workers to think that they are working nowhere or anywhere—in their specialty or profession, perhaps, or in 'cyberspace'.

V.   Nevertheless, their work will have a precise and practical influ-ence, first on the place where it is done, and then on every place where its products are used, on every place where its attitude toward its products is felt, on every place to which its by-products are carried.

VI.   There is, in short, no way of escape from the problems of effect and influence.

VII.   The responsibility of the worker is to confront these problems and deal justly with them. How is this possible?

VIII.   It is possible only if the worker knows and accepts the reality of the context of the work. The problems, the effect and influence, are inescapable because, whether acknowledged or not, work always has a context. Work must 'take place'. It takes place in a neighborhood and in a commonwealth.

IX.   What, therefore, must we have in mind when we go to work? If we go to work with the aim of working well, we must have a lot in mind. We must be mind-full. What, then, must we know? We can establish the curriculum by a series of questions:

X.   *Who are we?* That is, who are we as we approach the work in its inevitable place? Where are we from, what did we learn there, and (if we have left) why did we leave? What have we learned, starting perhaps with the influences that surrounded us before birth? What have we learned in school? More important, what have we learned *out* of school? What knowledge have we mastered? What skills? What tools? What affections, loyalties and allegiances have we formed? What do we bring to the work?

XI.   *Where are we?* What is this place in which we are preparing to do our work? What is the nature, what is the *genius*, of this place? What, if we weren't here, would nature be doing here? What will the nature of the place permit us to do here without exhausting either the place itself or the birthright of those who will arrive here later? What, even, might nature help us to do here? Under what conditions, imposed both by the genius of the place and the genius of our arts, might our work here be healthful and beautiful?

XII.   *What do we have*, in this place and in ourselves, that is good? *What do we need? What do we want?* How much of the good that is here, that we now have, are we willing to give up in order to have further goods that we need, that we think we need, or that we want?

XIII.   And so our curriculum of questions, revealing what we have in mind, brings us to the crisis of the modern world. Partly this crisis is confusion between needs and wants. Partly it is a crisis of rationality.

xiv.   The confusion of needs and wants is, of course, fundamental. And let us make no mistake here: this is an *educated* confusion. Modern education systems have pretty consciously encouraged young people to think of their wants as needs. And the schools have increasingly advertised education as a way of getting what one wants; so that now, by a fairly logical progression, universities are understood by politicians and university bureaucrats merely as servants of 'the economy'. And by 'the economy' they do not mean local households, livelihoods and landscapes; they mean the corporate economy. (If more and more of the powers that be think of education as merely the servant of the corporate economy, why should it be surprising that more and more of those same powers should think of the government as merely the servant of the corporate economy?)

xv.   But the idea that schools can have everything to do with the corporate economy and nothing to do with the health of their local watersheds and ecosystems and communities is a falsehood that has now run its course. It is a falsehood and nothing else.

xvi.   What actually *do* we need? We might say that, at a minimum, we need food, clothing and shelter. And, if we are wise, we might hasten to add that we don't want to live a minimal life; we would also count comfort, pleasure, health and beauty as necessities. And then, with the realization that it may be possible by reducing our needs to reduce our humanity, we may want to say also that we will need to remember our history; we will need to preserve teachings and artefacts from the past; we will need leisure to study and contemplate these things; we will need towns or cities, places of economic and cultural exchange; we will need clean air to breathe, clean water to drink, wholesome food to eat, a healthful countryside, places in which we can know the natural world—and so on.

XVII.   Well, now we see that in attempting to solve our problem we have run back into it. We have seen that in order to understand ourselves as fully human we have to define our necessities pretty broadly. How do we know when we have passed from need to wants, from necessity to frivolity?

XVIII.   That is an extremely difficult and troubling question, which is why it is also an extremely interesting question and one that we should not cease to ask. I can't answer it fully or confidently, but will only say in passing that our great modern error is the belief that we must invariably give up one thing in order to have another. It is possible, for instance, to find comfort, pleasure and beauty in food, clothing and shelter. It is possible to find pleasure and beauty and even 'recreation' in work. It is possible to have farms that do not waste and poison the natural world. It is possible to have productive forests that are not treated as 'crops'. It is possible to have cities that are ecologically, economically, socially, culturally and architecturally continuous with their landscapes. It is not invariably necessary to *travel* from a need to its satisfaction, or from one satisfaction to another.

XIX.   It is not invariably necessary to give up one good thing in order to have another. In our age of the world there is a kind of mind that is trying to be totally rational, which is in effect to say totally economic. This mind is now dominant. It is always telling us that the good things we have are really not as good as they seem, that they can seem good only to 'backward people', and are certainly not as good as the things we will have in the future, if only we will give up the things that seem good to us now. If a forest or a farm is destroyed to make a 'housing development', and the 'housing development' is then sacrificed to a factory or an airport, the rational mind wants us to believe that this course of changes is 'progress', and it offers as proof the successive increases in the value or the profitability of the land. It shows us the 'cost-benefit ratio'. And here we arrive at the crisis of rationality. We have come to the point at which reason fails.

xx.   Reason fails precisely in the inability of the cost-benefit ratio to include all the costs. We know that, however favorable may be the cost-benefit ratio, the progress from forest or farm to any sort of 'development' degrades or destroys the integrity of the local ecosystem and watershed, and we know that it causes human heartbreak. This kind of accounting excludes all coherences except its own, and it excludes affection. The cost-benefit ratio is limited to what is handily quantifiable, namely money. The failure of reason comes to light in the recognition that things which cannot be quantified—the health of watersheds, the integrity of ecosystems, the wholeness of human hearts—ultimately affect the durability, availability and affordability of necessary quantities. To think of landscapes merely in terms of economic value will in the long run reduce their economic value, not to mention the availability of such necessities as timber and food, clean water and clean air.

xxi.   The mind makes itself totally rational in an effort to become totally comfortable, but at the risk of eventually becoming totally uncomfortable. The cost of subordinating all value to economic value will eventually be economic failure.

xxii.   We are well acquainted with this mind of would-be total rationality, hellbent on quantification. And we are increasingly well acquainted with its results in the ruin of culture and nature. And so the next in our curriculum of questions necessarily is this: Do we know of a different or better or saner kind of mind?

xxiii.   I think we do. It is what I would call the affectionate or sympathetic mind. This mind is not irrational, but neither is it primarily rational. It is a mind less comfortable than the mind that aspires only to reason, and it is more difficult to define.

xxiv.   It is defined, I think, in the parable of the lost sheep in the Gospels of Matthew and Luke, and in the Buddhist vow: 'Sentient beings are numberless, I vow to save them'. The mind given over to reason would lose no time in demonstrating mathematically that it 'makes no sense' to leave ninety-nine sheep perhaps in danger while you go to look for only one that is lost. And surely it makes even less 'sense' to vow to save all sentient beings.

xxv. Obviously, to assent to such teachings as these we must change our minds. We must give up some part of our allegiance to reason and to quantification, and we must accept as our lot in life a perhaps irreducible discomfort. We have given affection and sympathy a priority over rationality. We have consented to the proposition that at least a part of what we have now, the part we have been given, is good, and we have assumed the responsibility of preserving what is good. We have assented implicitly to God's approval of His work on Creation's seventh day.

xxvi. To change one's mind in this way is to change the way one works. This changed way of working is new to us in our industrial age, but is old in the history of human making. And what is this way? How does this changed mind go about its work?

xxvii. Such a mind, I think, is no longer satisfied with the conventional standards of industrialism: profitability and utility. Needing a more authentic, more comprehensive criterion, it looks beyond those concerns, without necessarily giving them up. It tries to see the work and the product in context; it tries to derive its standards from that context. And once again it must proceed by way of questions. Is the worker diminished or in any other way abused by this work? What is the effect of the work upon the place, its ecosystem, its watershed, its atmosphere, its community? What is the effect of the product upon its user and the place where it is used?

xxviii. Work under the discipline of such questions might hope to give us, to name a few examples, forestry that would not destroy the forest, farming that would not destroy the land, houses that would be suited to their places in the landscape, products of all kinds that would neither exhaust their sources nor degrade their users.

XXIX.   Obviously, there has come to be a radical disconnection between the arts and sciences and their ultimate context, which is
always the natural or the given world. Why should this be?

XXX.   I venture to think that it is a problem of perception, most particularly and directly in the sciences. The scientific need for
predictability or replicability forces perception into abstraction. The
'test plot', for example, is perceived, not as itself, but as a plot *representative* of all plots everywhere.

XXXI.   Developers of technology, in somewhat the same way, are
under commercial pressure for *general* applicability. The place where
a new machine or chemical or technique is proved workable is
assumed to be *representative* of all places where it might work.

XXXII.   These processes in science and technology seem to be closely
parallel in effect to the processes of centralization in economic and
political power.

XXXIII.   The result is that all landscapes, and the people and other
creatures in them, are being manipulated for profit by people who
can neither see them in their particularity nor care particularly about
them.

XXXIV.   The disciplines that are not directly involved in this
manipulation have nonetheless consented to it. It is the problem of
all the disciplines.

XXXV.   It seems to me that the solution to this problem is not now
foreseeable, but I believe it can come about only by widening the
context of all intellectual work and teaching—perhaps to the width
of the local landscape.

XXXVI.   To accept so wide a context, the disciplines would have to move away from strict or exclusive professionalism. This does not imply giving up professional competence or professional standards, which have their place; but professionalism as we now understand it has already shown itself to be inadequate to a wide context. To bring local landscapes within what Wes Jackson calls 'the boundary of consideration', professional people of all sorts will have to feel the emotions and take the risks of amateurism. They will have to get out of their fields, so to speak, and into the countryside and the city and the community; and they will have to be actuated by affection.

XXXVII.   In the sciences, I think the acceptance of the local landscape as context will end the era of scientific heroism. No one scientist or one team of scientists or one science-exploiting corporation can expect to 'save the world', once the disciplines have accepted this context that is at once wide and local. The solutions then will have to be local, and there will have to be a myriad of them. The scientists, moreover, will have to suffer the responsibility of applying their knowledge at home, sharing the fate of the place where their knowledge is applied.

XXXVIII.   Throughout these notes I have been assuming—as my reading and the work I have done have taught me to assume—that it is impossible for us humans to know in any complete or final way what we are doing.

XXXIX.   Now I will explain this assumption in a different way, but one that leads to the same conclusion.

XL.   Increasingly since the Renaissance, the building blocks of rational thought have been facts, pieces of data that can be proved or demonstrated or observed to be 'true'. So great has been our confidence in facts, and in the empirical processes by which factuality is tested, that Thomas Jefferson, for example, could speak smugly of 'our barbarous ancestors'.

XLI.   The assumption seems to be that the pursuit of truth in our time, as never in the times before, has become completely scientific and rational, so that we now not only possess more facts every day than we ever did before, but have only to continue to fill in the gaps between facts by the empirical processes of our science until we will know the ultimate and entire truth.

XLII.   I do not believe this. I think it is a kind of folly to assume that new knowledge is necessarily truer than old knowledge, or that empirical truth is truer than non-empirical truth. But I also do not believe that factual truth is or ever can be sufficient truth, let alone ultimate truth.

XLIII.   A fact, I assume, is not a thing, but is something known about a thing. H_2O is a fact about water; it is not water. A person who had never seen water could not recognize it, much less recognize ice or steam, from knowing that formula. Recognition would require knowledge of many more facts. Water is water because it is the absolute sum of all the facts about itself, and it would be itself whether or not humans knew all the facts.

XLIV.   The only true representation of a thing, we can say, is the thing itself. This is true also of a person. It is true of a place. It is true of the world and all its creatures. The only true picture of Reality is Reality itself.

XLV.   In order to work, in order to live, we humans necessarily make what we might call pictures of our world, of our places, of ourselves and one another. But these pictures are artefacts, human-made. And we can make them only by selection, choosing some things to put in the picture, and leaving out all the rest.

XLVI.   From the standpoint of the person, place or thing itself, of Reality itself, it doesn't make any difference whether our pictures are factual or imagined, made by science or by art or by both. All of them literally are fictions—things made by humans, things never equal to the reality they are about and never assuredly even adequate to the reality they are about.

XLVII.   Facts in isolation are false. The more isolated a fact or a set of facts is, the more false it is. A fact is true in the absolute sense only in association with all facts. This is why the departmentalization of knowledge in our universities is fundamentally wrong.

XLVIII.   Because our pictures of realities, and of Reality, are invariably and inescapably incomplete, they are always to some degree false and misleading. If they become too selective, if they exclude too much (on the ground, for instance, of being 'not factual'), if they are too biased, they become dangerous. They are constantly subject to correction—by new facts, of course, but also by experience, by intuition, and by faith.

XLIX.   We may say, then, that our sciences and arts owe a certain courtesy to Reality, and that this courtesy can be enacted only by humility, reverence, propriety of scale, and good workmanship.

VANDANA SHIVA
Annadāna – Gift of Food
☙

Today is 16 October, a day celebrated as World Food Day.
When the Prince's Foundation wrote and asked me to be here
on the 16th, I accepted the generous invitation even though it is
usually a time when I would be back home organizing something or
other around the food issue. I thought that this year I would dedi-
cate World Food Day to a reflection on the ecology of food, as part of
this series entitled *Ecology: A Sacred Trust*. I wanted to explore how all
that we have learned about respecting food, venerating food, producing it safely and sustainably, ensuring that everyone has a right
to food, is being erased by a new commoditization of culture that is
giving us an economy in which species are being wiped out, small
farmers are being wiped out, and our health is being wiped out.

I think that the first thing to recognise about food is that it is the
very basis of life, and this is something which ecologists very often
forget. They treat food as one thing and Nature as wilderness somewhere else: if you are producing food you cannot have Nature, if you
have Nature you cannot meet human needs. And so we have built up
these amazing dualisms that force us into constantly more and more
destructive routes toward meeting our vital needs, fooling us into
believing that the more resources you consume and destroy through
intensive agriculture the more you 'save' Nature. But food is not just
our vital need, food is the basis of being. As the Taittirīya Upaniṣad says:

From food (*anna*), verily, creatures are produced
Whatsoever [creatures] dwell on the earth . . ..
For truly, food is the chief of beings.[11]

Beings here are born from food, when born they live by food,
on deceasing they enter into food.[12]

Food is alive: it is not just pieces of carbohydrate, protein and nutrient, it is a being, it is a sacred being.

Verily, they obtain all food
Who worship Brahma as food.[13]

This entire Upaniṣad is dedicated to the giving of food: if someone was to ask me to name the one text in the world that is about the ecology of food as a sacred trust, I would say, 'Just read the Taittirīya Upaniṣad'.

Not only is food sacred, not only is it living, but it is the Creator Itself, and that is why in the poorest of Indian huts you find the little earthen chula, or stove, being worshipped; the first piece of chapati set out for the cow, the next piece for the dog, then you find out who else is hungry in your domain. In the words of the Mahā Aśvamedhikā:

The giver of the food is the giver of life, and indeed of everything else. Therefore, one who desires wellbeing in this world and beyond should specially endeavour to give food . . ..
Food is indeed the preserver of life and food is the source of procreation.[14]

Because food is the very basis of creation, food is creation, and it is the Creator. It is Divinity in the context of the way we live: there are all kinds of duties that we should be performing with respect to it. If people have food it is because society has not forgotten those duties. If people are going hungry, society has moved away from the ethical duties related to food.

The very possibility of our being here, the very possibility of our living, is based on the lives of all kinds of beings that have gone before us—our parents, our mothers, the soil, the earthworm—and that is why the giving of food in Indian thought has been treated as the everyday yajña, or 'sacrifice', that you have to perform. It is not the once-on-a-Sunday ritual, it is a ritual embodied in every meal, every day, all of the time, reflecting the recognition that giving is the condition of your very being. You do not give as an extra, you

give because of your interdependence with all of life, your interdependence both with the human beings who make your life possible in your community, and with the non-human kith and family
that we have. One of my favourite images in India is the kolam, a
design which a woman makes in front of her house. On the days
of Pongal, which is the rice harvest festival in South India, I have
seen women get up at three in the morning to make the most
beautiful art work outside their huts, and it is always made with
rice. The real reason they did that was to feed the ants, but it ended
up being such a beautiful art form that it has just gone on from
mother to daughter, and at Pongal time you actually have everyone
trying to do the best for their offering, which in this case is really an
offering to ants, in the most amazing beauty. Women will start at
three or four o'clock in the morning and go on until six or seven,
and for two or three hours they will just be making these beautiful
designs, never stopping laying it out.

We do a lot of conservation work, conservation of seeds. There are
two kinds of rices, the japonica and the indica. The Indian rice
varieties are the ones that give you those beautiful grains of pilau
and biriyani. Because of the way the food culture has evolved in
Japan and China, where they like to eat with chopsticks, they have
selected the sticky rices and the more thick-grained rice, not the
ones in which each grain cooks separately, so that they can lift it with
their chopsticks. But since we eat with our fingers, and do not need
rice grains to hang together, we can manage the separate grains. The
indica rice variety's homeland is a tribal area called Chattisgarh in
India. It must be about fifteen years ago that I first went there. They
weave beautiful designs of paddy, which they then hang around the
huts. I thought that this was related to a particular festival, and I
said, 'What festival is it for?' And they said 'No, no, this is for the
season when the birds cannot get rice grain in the field.' They were
ensuring that they were putting rice out again for other species, in
very beautiful offerings of art work. I have now seen that kind of
grain being sold in gift stores at Thanksgiving and Christmas,
because it has become reduced to something that is totally separate

from its ecological roots, and is merely an item of decoration. Why those beautiful designs of grain were made has been forgotten.

Because there is this amazing owing of the conditions of our life to all other beings and all other creatures, giving—to humans and to non-human species—has been part of this amazing gift of food. The one thing Satish Kumar and I have been trained in among our basic norms and basic values in Indian culture is *annadāna*, the giving and the gift of food. All other ethical arrangements in society get looked after if everyone is engaging in *annadāna* on a daily basis. According to an ancient Indian saying: 'There is no *dāna* greater than *annadāna* and *tīrthadāna*, the giving of food to the hungry and water to the thirsty'. Or again, in the words of the Taittirīya Brāhmaṇa:

> Do not send away anyone who comes to your door without offering him food and hospitality. This is the inviolable discipline of humankind; therefore have a great abundance of food and exert all your efforts towards ensuring such abundance, and announce to the world that this abundance of food is ready to be partaken by all.[15]

Thus from the culture of giving you have the conditions of abundance, and the sharing by all. That is why your pots must keep overflowing, that is why the bells in your village must ring to say 'Is no one hungry anymore?' And we have scripture after scripture saying that if you eat before the last person in your sphere, or while the last being in your sphere of influence is still hungry, then you are already committing sin, because you should only eat after everyone's needs have been met.

The violation of that ethic is in my view the beginning of all nonsustainability and all injustice. I will try to work through how our current situation of more and more people being denied their basic needs, the basic possibility of engaging in creative productive activity, and indeed of getting enough food, is linked to our strange search for a very false sense of growth by alienating the sacred basis of food, and turning it into a mere commodity.

In another scripture it is said:

I forsake the one that eats without giving. I am the *annadevatā*
(the god of food, the divine in food): I come and go according
to my own discipline. I nurture the one for whom giving
carries the same significance as eating. To him I reach in
plenty; I remain out of reach of the other who eats without
giving. Who amongst men can deter me, the *annadevatā*, from
my course?

If you really look at what is happening in the world, we have so-called
surpluses growing with hunger. We have more and more food in
the world, while 820 million people go hungry. As an ecologist,
I see these surpluses as pseudo-surpluses. They are pseudo-surpluses
because the overflowing stocks and packed supermarket shelves are
the result of production and distribution systems which take food
away from the weak and marginalized, and from other beings. I do
not believe that we have more food in the world. We do have more
food in our supermarkets—I had to pick up something the other day
and go through the basement of Marks & Spencer, and I just went
dizzy, immediately, just seeing the food there, because I could see
how that last peasant's rice field has been converted into a banana
plantation to get luscious bananas here. The desecration of coastal
farms and fishermen's livelihoods has brought the shrimps to
Northern markets. Each time I see supermarkets, I see how every
community's, every ecosystem's capacity to meet its food needs is
being undermined, so that a few supermarkets, and a few people in
the world, can have an appearance and experience of surpluses. But
these are pseudo-surpluses, because they are not bringing you more
nourishment, and that is why we have 820 million people in the
world, which is one quarter of humanity, not having enough to lead
a healthy life, and the rest of the people eating the foods that are
produced by this rush for growth in surplus getting bad food and
getting ill anyway. For the latter are those who eat before giving,
without ensuring that the last person has been liberated from
hunger: in that context, even the well-fed have diseases. So you have
literally the whole world suffering from diseases of the denial of

nourishment: those who do not have food, and those who have too much of it. For the very way in which this food is being produced, processed and distributed is violating the sacred trust which recognizes food as the basis of life.

Let us begin with how it is produced. Part of what anyone with a basic knowledge of the ecology of food production and the ecology of soils knows is that to have sustainable food supplies we need our soils as living systems: we need all those millions of soil organisms that make fertility. And that fertility gives us healthy foods. In industrial culture we thought that it is not the earthworm that creates soil fertility, we thought that soil fertility comes from the war-surpluses of explosives factories; that pest-control does not come out of the balance of different crops hosting different species but from poisons. When you have the right balance, they never become pests: they all exist, and none of them destroys your crop. As you reduce that diversity, as you reduce the multiplicity of plants, you also create an epidemic of pests because you destroy the homes and habitats of the diverse species; and before you know it you need to start spraying pesticides, which are also a conversion from war, and you say that the chemical fertilizer and the pesticides are going to give you more food.

The recently released report of the Food and Agriculture Organization has chart after chart of bubbles, growing bigger and bigger, to show how in the last century we have had so much increase in food productivity: you would really think we had more food. All that they have calculated is labour displacement. They have only looked at labour productivity, at how much food a human being produces by technologies that are labour-displacing, species-displacing and resource-destroying. It does not mean that you have more food per acre; it does not mean that you have more food per unit used of water; it does not mean that you have more food for all the species that need the food, because our earthworms need the food, our cows and cattle that give us organic manure and renewable energies need the food. All of these diverse needs are getting erased as we define productivity on the basis of how much labour we can displace from the land, in the fastest way. That does not mean more food in the world; it definitely means less food because all those heavy machines are just chopping up your soil, chopping up the earthworms. Now usually this is happening underground and we are not seeing the violence we are doing to all the beings that are the condition for our producing food in sustainable ways. No one has ever calculated how much less food the organisms are getting, how much less food the earthworm is getting, how much less food the cows are getting, how much less food the birds are getting.

In fact we are now working out very smart technologies, based on genetic engineering, which accelerate the logic of war with other beings. On my recent trip to Punjab, it suddenly hit me: they no longer have pollinators. And so we have started a whole new movement in the scientific community, and among students. The students are wonderfully sensitive: you just tell the children about pollinators and they make you the most beautiful stories about butterflies and bees. For them these creatures are vividly alive. It is quite otherwise with those amazing people who are manipulating crops to put genes from the bt-toxin (the soil bacterium *bacillus thuringensis*) into plants, so that the plant is releasing toxins at every moment and in every cell: its leaves, its roots, its pollen. These toxins are then being eaten by the ladybirds, which are dying; by the butterflies, which are dying. There was a very famous Cornell study of the monarch butterfly, showing how, with these new genetically engineered crops, we are creating poison factories in our plants, while thinking we have achieved something brilliant in the process.

It is only brilliant to the extent that we do not see the web of life that we are rupturing, because we are not aware of the web of life. You can only see the interconnections to the extent that you are sensitive to them, to the extent that you are aware of them. And when you are aware of them you immediately recognize what you owe to other beings, to the farmers who have produced the food, and the people who have nourished you when you could not nourish yourself, those amazing years at the beginning of your life and at the end of your life when you need others to nourish you. And the giving

of food, basically the sacred trust about food, is related to the idea
that every one of us is born, as we say, in ṛṇa. We are born in ṛṇa or in
debt to other beings: our very condition of being born depends on
this debt. So we come with a debt and for the rest of our lives we are
paying back that debt—to the bees and the butterflies that pollinate
our crops, to the earthworms and the fungi and the microbes and the
bacteria in the soil that are constantly working away to create the
fertility that our chemical fertilizers can never, never replenish. They
can only pollute: they can contaminate, they can kill, they can kill
sources of life that we have stopped noticing, but which are
absolutely essential for the maintenance of productivity.

Man, therefore, is born and lives in debt, in ṛṇa, to all Creation,
and it becomes our duty to recognize this. The gift of food, annadāna,
is merely a recognition of the need for constantly paying back that
obligation, that responsibility; it is merely a matter of accepting
and endeavouring to repay our debts to Creation, and to the communities of which we are a part, that we have incurred not by doing
good in our lives but by just arriving in this world. It is merely
being humanly responsible. And that is why again most cultures
that have seen ecology as a sacred trust, and our maintenance of our
life's conditions as a maintenance of that sacred trust, have always
spoken of duty or responsibility, of dharma. Rights have flowed out
of responsibility, because once I am ensuring that everyone in my
sphere of influence is fed, someone in that sphere is also ensuring
that I am fed.

When I left university teaching in 1982, everyone said, 'How will
you manage without a salary?' I replied by saying that if 90% of India
manages without a salary, all I have to do is put my life in the kind
of relationships of trust that they live through. If you give then you
will receive. You do not have to calculate the receiving, what you
have to be conscious of is the giving. I actually have quite nice saris,
my collection of saris is all gifts; this one my sister gave, something
else my brother will give, and you look after each other like that.
Indians do not buy clothes for themselves, they usually receive
festival gifts on Durga Puja, on Diwali, and your needs are looked after

through all that flow of giving, of *dāna*, just as in the case of food the calculus is of giving on the basis of the debt, of the ecological debts, that we have.

What is the context now? We also have debts, except that now they are financial debts, so that a child born in India or in any Third World country already has millions of dollars of debt on its head owed to the World Bank, which then has every power in the world to tell you and your country that you should not be producing food for the earthworms and the birds and the animals and the cows, or for the people of the land: you should be growing shrimps and flowers for export, because that earns money. It does not earn very much money, either. I have made calculations that show that these export-oriented systems generate one dollar of trade for international business but bring nothing but destruction back to the peasant who was displaced. To replace a peasant farm growing millet with a shrimp factory means that the peasants are wiped out in the process; so they are not among the people who earn out of this, they lose. A dollar of trading by international business, in terms of profits, is leading to $10 of ecological and economic destruction in local ecosystems: livelihoods of farmers, production of fish, production of manure, production of forest goods, production of food. Now if for every dollar being traded we have a $10 shadow-cost, in terms of how we are literally robbing food from those who need it most, you can understand why, as growth happens and as international trade becomes more intense, you naturally and inevitably have more hunger, because the people who needed that food most are the ones who are being denied access to food by destruction of these livelihoods and local food systems, by this new system of trading. This so-called free trade is taking away from them any way of looking after others' needs, or their own.

We can think on three levels: the ecological level, based on the rights of all species; the productive level, based on food quantity for human needs; and the distributive level, based on how food is exchanged and traded. At the ecological level, we must think in terms of the web of life, which in a sacred trust gives you the obligations of *dāna*, gives you the obligation to give food, the gift of food. But human needs are also met through the sacred trust which you build into the maintenance of life systems, so that sustainability comes as a by-product of living rightly, in the ethical sense, in the sense of spirituality; and also as a result of that, if you've taken care of all the beautiful beings in the soil, you will get nourishing food. The food that is produced by poisoning our soils also poisons our bodies. There is a chapter in my book *Tomorrow's Biodiversity* which answers a question posed by the publisher: how can we protect biodiversity if we are to meet growing human needs?[16] My reply is that the only way to meet growing human needs is to protect biodiversity, because unless we are looking after the earthworms and the birds and the butterflies we are not going to be able to look after people either. And this idea that somehow the human species can only meet its needs by wiping out all other species is just a wrong assumption, it is based on not seeing how the web of life connects us all, and how much we live in interaction and in mutual interdependence.

At the level of how much food we are producing for humans, we have worked out and demonstrated, on the basis of biodiversity calculus for farm after farm, that the idea that industrial mono-cultures somehow produce more food is just not true, even in purely quantitative terms. Monocultures produce more monocultures, but they do not produce more nutrition. If you take a field and plant it with twenty crops, it will have a lot of food output, but if any one of those individual yields—say of corn or wheat—is measured in comparison with that of a monoculture field, of course you will have less, because the field is not all corn. So just by shifting from a diversity-based system which respects diverse species into a monoculture industrially supported with chemicals and machines and everything else that goes with it, you automatically, tautologically, define it as more even though you are getting less—less species, less output, less nutrition, less farmers, less food, less nourishment. And yet we have been absolutely brain-washed into believing that when we are producing less we are producing more. It is an illusion of the deepest kind.

At the level of distribution and trade, the final kind of calculus, it is not the case that people did not trade before. Right in my region— I come from the Himalaya—people have for centuries traded the wool which they produce, or the amaranth seeds which they grow, for the salt which the plains people can bring them. The only two things that they get from the plains would be salt and oil, and in the high-altitude villages those are also still the only things they depend on; everything else they produce themselves. Trade today is no longer about the exchange of things which we need and which we cannot produce ourselves. Trade is an obligation to stop producing what you need, to stop looking after each other, and to buy from somewhere else. But if it is a system of buying from somewhere else which we have been coercively and violently locked into by the WTO, by the World Bank and IMF, it is not mutually independent companies, people, making ethical choices, engaging in ecologically responsible action, *qua* buying and selling; the very nature of this kind of buying and selling and globalized trade ends up concentrating power in the hands of three or four players.

In trade today there are four grain giants. The biggest of them, Cargill, controls 70% of the food traded in the world; and they fix the prices. They sell the input, they tell the farmer to grow something, they buy back cheap from the farmer, then they sell it at high cost to consumers. In the process they poison every bit of the food chain; in the process, instead of giving, they are thinking of how they can make the next efficient technology, and the next efficient economic arrangement, that can take out that last bit, from eco-systems, other species, the poor, the Third World. You might recall that in the Reith Lecture I quoted two responses that had been made in my presence, one by Cargill and one by Monsanto, when we were talking about seeds and the failure of seeds—which, as you will know from my writings as well as from a recent public hearing which I have organized, has actually wiped out 20,000 farmers who committed suicide.[17] But in the course of this whole seed debate in the early 90s Cargill said, 'Oh, these Indian peasants are so stupid that they do not realise that our seeds are so smart: we have found new technologies that prevent the bees from usurping the pollen'. Now the concept of 'the gift of food' tells us that pollen is the gift that we must maintain for pollinators, and therefore we must grow open pollinated crops, that bees and butterflies can pollinate. That is their food, that is their food chain, and it is their ecological space. And we merely have to make sure that we do not eat into their space. Instead, Cargill says that the bees usurp the pollen because they have defined every piece of pollen as their property, so that they can design the crops that do not rejuvenate freely, and own and patent these crops. And in a similar way, in a debate on these herbicide-resistant genetically engineered varieties called Roundup Ready, varieties that wipe out the diversity of our crops, Monsanto said: 'Through use of Roundup we are preventing weeds'—which in our context are sources of nutrition—'from stealing the sunshine'. The entire planet is supposed to be energized by the life-giving force of the sun. And now Monsanto has basically said: 'No, it is Monsanto and the farmers in contract with Monsanto that, alone on the planet, have a right to the sunshine, the rest of it is theft'.

So what you are getting is a world which is absolutely the obverse of the world of the giving of food, because it is the sacred trust. Instead, it is the taking of food from the food chain and the web of life, and that is where most technologies are progressing: from the hybrid to the terminator, to all the so-called 'smart technologies', which are just cleverer at taking away and therefore apparently producing more by denying other beings their rightful share of the gifts of the planet, and the gifts of the cosmos. So instead of *dāna*, we have profits and greed as the highest organizing principle. Unfortunately, the more the profit, the more the hunger, the more the illness; the more the destruction of Nature, of soil, of water, of biodiversity, the more non-sustainable our food systems become; and we then actually become surrounded by deepening *adāna* ('non-giving'), which is showing up in terms of more suicides of farmers, or of farmers selling their kidneys to pay back debts. Not the ecological debt to Nature, to the earth and to other species, but the financial debt to the money-lenders and to the agents of chemicals and seeds. That ecological debt is in fact replaced by this financial debt: the giving of nourishment and food is replaced by the taking of more and more in terms of profits.

This is, in my view, a deep climate in which we are all being forced, against our will, to participate in very pervasive sin. And I think that what we need to do now is to start to find ways of detaching ourselves from these sinful arrangements, because it is nothing less than that. It is not just replacing free trade with fair trade, it is not little bits and pieces: unless we see how the whole is leading to a poisoning and pollution of our very beings, of our very consciousness, we will not be able to make the deeper shifts that allow us to create abundance again, because in the taking of more we are not creating abundance, we are creating scarcity. The growing hunger is part of the scarcity. And the growing diseases of affluence are a third part of the scarcity: we are actually not producing abundance because we are basing everything on 'How can you take more?'—how can you take more out of Nature, how can you take more out of the Third World, how can you take more out of the poor?

If we relocate ourselves again in the sacred trust that is ecology, and recognize our debt and duty to all human beings and non-human beings, the protection of the rights of all human and non-human species simply becomes part of our ethical norm and our ethical duty. And the care of the non-human leads then to the care of the human, it leads to our being able to maintain our farmers on the land. If we realize that our farmers are basically earthworm-carers we can be absolutely sure that the combine-harvesters are the wrong thing to have—we do need people, who replenish the soil. And as a result of that, those who depend on others for feeding them and for bringing them food, the consumers, will also get the right kind of food and the right kind of nourishment. The care of all creatures means the care of our cultivators; and the care of our cultivators means the care of our consumers. So if we begin with the nourishment of the web of life, we actually solve the economic crisis of small farms, the health crisis that consumers are facing, and the economic crisis of Third World hunger and Third World poverty—all these in a way just flow out of relocating our dependence on other life-forms, and ensuring that their food needs are the basis of how we grow and produce food.

SATISH KUMAR
Reverence for Life: A Jain Perspective
❧

In this series of lectures with the theme Ecology: A Sacred Trust, I am going to talk about the Jain perspective. There are three main religious traditions in India that come from a single source. They are the so-called Hindu, Buddhist and Jain traditions. I say 'so-called Hindu' because there is no such thing as 'Hindu'. No such word exists in the ancient Indian texts. The word 'Hindu' is a Western invention and related to the 'Indus' river. For when Western people came to that part of India, they called the people who lived beyond the Indus 'Hindus'.

The reason why Western scholars called everybody from the subcontinent 'Hindus' was because there was no one religion there. In India there is no one god. There is no one prophet. There is no one book. There are thousands of gods, and thousands of religions. There is no one 'Hindu' religion. But Western scholars like to classify. So the people beyond the Indus were classified as 'Hindus'.

Then those who follow the teachings of the Buddha are Buddhist, and those who follow the teachings of Mahāvīra are the Jains. Mahāvīra was called Jina, which means the conqueror of inner enemies such as ego, pride, anger and delusion. Thus followers of Jina became 'Jains'. Here too there is some controversy among the historians and scholars. Some believe that Buddha and Mahāvīra are the same person. They were born in the same part of India and at the same time. Their teachings are also very similar. But among their disciples there was a great schism. As a result the two separate orders were established. That is how Buddhism and Jainism came into being. Here we will not go into these controversies. Suffice it to say that the three great traditions of India have much in common. Even though I am going to talk about the Jain perspective, my talk should be appreciated in the broader context of Indian traditions as a whole.

Mahāvīra, born as a prince, simply realized that there is more to life than the endless pursuit of happiness through fame, fortune and power. In fact fame, fortune and power are the bringers of 'unhappiness'. This notion was not based in any intellectual or philosophical theories, rather it was based in his own experience. As a man in his position, he already had enough of fame, fortune and power. Also of course he was expected to enlarge his base of 'fame, fortune and power'; but he found that such pursuit only brought him anguish, pain, fear and dissatisfaction. Mahāvīra decided not to follow the path of 'fame, fortune and power'. He left the kingdom for the forest in search of wisdom, liberation and enlightenment. After a long life of meditation and contemplation, Mahāvīra came to the conclusion that there are three essential elements which sustain life: he called them *ahiṃsā* (non-violence), *saṃyama* (simplicity) and *tapas* (spiritual practice).

Ahiṃsā, or 'Non-violence'

Mahāvīra said that non-violence must begin in our mind. Unless our mind is calm and compassionate, non-violence is not possible. Unless we are at ease in our inner world, we cannot practise non-violence in the external world. If our mind is condemning other people yet we speak sweet words, then that is not non-violence. The seeds of non-violence live within our consciousness. Keeping our consciousness pure is therefore an essential part of Jain non-violence. Pure consciousness means consciousness that is uncorrupted, uncontaminated and undiluted with the desire of controlling others.

Why are we violent? Because we wish to control people, we wish to control animals, we wish to subjugate Nature. We say: 'We own everything here. This world is for our benefit.' Those who eat meat say: 'Animals and fish are there for our food.' Everything is for us and we are the masters of others. This is the violence of the mind, which leads to physical violence.

Mahāvīra saw the world as a sacred place. The birds, the flowers, the butterflies and the trees are sacred. The river is sacred, the soil is sacred and the oceans are sacred. Life as a whole is sacred. All our interactions with other people and with the natural world must be based on this sacred trust. A deep reverence for life.

Non-violence of the mind should be translated into non-violence of speech. Harmful, harsh, untrue, unnecessary, unpleasant and offensive speech can be lethal. Skilful use of language is a sacred skill. Mahāvīra insisted that we must understand others fully before we speak. If we have not understood what the other people are saying then with patience we should inquire, explore and listen, and only then speak the truth. Language can only express partial truth, therefore non-violence is an essential guide to our spoken words. According to Mahāvīra we have one mouth and two ears. So we must practice two parts listening and one part speaking. In this context Mahāvīra put a very high value on silence. A Jain monk is called *muni*, which means 'the silent one'.

Non-violence of mind and speech should be further translated into the non-violence of action. Ends cannot justify the means. The means must be compatible with the ends. Therefore all our actions must be friendly, compassionate and unaggressive. Do no harm. No ifs and no buts. No compromise. Mahāvīra's non-violence is unconditional love to all beings.

Thus non-violence is the paramount principle of the Jains. All other principles stem from non-violence. Non-violence is first and last. Because all life is sacred we may not violate or take advantage of those life forms which may be weaker than ourselves.

Samyama, or 'Simplicity'

The second principle is *samyama*, which means 'simplicity'; this has also been translated as 'self-restraint', 'sufficiency' or 'frugality'.

It seemed to Mahāvīra that human beings have a tendency to expand their territory. For example, we ask 'Is your business expanding?' The idea of economic growth has become a new religion. For the last one hundred years we have been worshiping the god of economic growth. That is totally opposite to the Jain principle of *samyama*.

Being satisfied with less is *samyama*. The idea that whatever we have, or however much we have, is never enough is the source of anguish. We need to move from 'more and more' to 'enough!'.

There is nothing lacking in the world. Wherever we look we can see the abundance of Nature. But when we want to own, control and possess it, we create scarcity. Because we can never possess everything, we always want more. This possessiveness is the source of scarcity. The moment we are satisfied, and don't want to control and possess, we have the whole world at our disposal. When we are at the service of the world, the world is also at our service. This is complementarity, reciprocity, mutuality, interdependence and interconnectedness.

When we are able to build our civilization on the principle of *samyama*, then we are able to take pleasure in the small and simple things of life. My mother once explained the meaning of *samyama* to me. She said, 'If you have too many possessions, if you have too many things to do, if you have too many places to go, if you have too many clothes to wear and too many pots and pans to wash, all your time is spent on taking care of material things. When do you have time to meditate, to take care of your soul?'

Now that is a very beautiful way to look at life. If we have a few things, we can learn to appreciate them and take care of them, and they will last for a long time. And then we can learn to make beautiful things. The Jains wish to have few things, but they learn to appreciate them and look after them. That is the way of *samyama*.

The material things are not there simply to decorate us, or merely to be useful to us; we develop a relationship with them. We have a relationship with a chair, a table, pots and pans, with the clothes that we wear. We even develop a friendship with our clothes! We don't own things to look slick and smart or to impress others. Rather, we relate to things and value things as sacred objects. As in *ahiṃsā* so in *samyama*: not only is life sacred, but even matter is sacred.

So the material and the spiritual become parts of a continuum. No dualism. When we are trying to possess, control and own, rather than to take care, then things become a burden. The mentality of control leads to materialism. If you ask 'What is the difference between a materialistic and a spiritual view of life?', Mahāvīra would say that the spiritual view of life looks at the world with a sacred eye. This is the 'third eye'. This eye is the eye of the heart, the eye of imagination, the eye of the soul, the eye of the spirit. When we look at a chair with two eyes we see only a three-dimensional object, a useful piece of furniture on which we can sit. But when you look with the third eye, it is a sacred object, it is a gift. We think of the tree from which it was made. We think of the craftsperson who made it. We develop friendship, relationship, respect, gratitude and reverence. Even for a chair, Jains have reverence.

Tapas, or 'The Spiritual Practice of Purification, Sacrifice and Fasting'

The principle of tapas (literally 'heat') or purification is the most difficult to understand. But understand we must, because it is a very important part of Mahāvīra's thought. During our daily life our bodies get sweaty and smelly. So we purify our bodies by having a shower. Our clothes get dirty, so we wash them. Our houses get dirty, so we sweep the floor, even scrub it and remove the cobwebs. There is a tradition of spring-cleaning. These practices help to purify our external environment. But what about the internal environment? Don't we also need practices for inner purification? Our minds get polluted with harmful and 'dirty' information. Our consciousness gets contaminated by the dirt of ego, greed, pride, anger and fear. Our souls get 'sweaty and smelly' with desires, attachments and anguish. So Mahāvīra devised ways to purify the minds, the souls and the consciousness, and he called this tapas.

Take my sister, for instance. Last year she was in a tapas, of fasting. In such a period, she will eat food for one day and fast the next. Then eat for one day and fast for two days. Then eat for one day and fast for three days, and then four days, and five days. Up to fifteen days she will fast. And then she will reverse the process and fast for fourteen days, eat for one day and fast for thirteen days, eat for one day and fast for twelve days, and so on back to one. This is a practice of going through the heat of austerity, a kind of bonfire of ego, pride and greed.

For many years my mother would never eat on two consecutive days. She would always eat on one day and fast the next. At the times of the celebration of religious festivals, she would not eat for eight days. When she was eighty, she said, 'Now I am eighty. I have lived long enough. I do not need to carry on.' And so she decided to embrace death. She went to her family and to everyone that she knew, saying goodbye. And then she said, 'From tomorrow I am going to fast unto death.' And so she did!

Now just imagine that sort of idea in our modern world. People would think that it was crazy, they would even say that it is suicide!

But death is not something to be frightened of. We are all going to die. My mother would say, 'Why not embrace death rather than be frightened of it and run away from it?' The ultimate sacrifice is the sacrifice of our own body at the right time, in the right spirit and in the right context.

Fasting, meditation, restraint, pilgrimage and service to others fall within the category of *tapas*. It is a kind of soul exercise to keep our inner world healthy, fit, clean and pure. Just as people go to a gym, or jog, or go for a daily walk to keep their bodies fit, so we also need to do yoga, meditation, fasting and service to the community to keep our souls fit. That is *tapas*.

I have tried to explain these three teachings of Mahāvīra as simply as possible. Of course there are more scholarly and more academic explanations of these teachings. To find such explanations, further studies of the original scriptures should be undertaken.[18] Ultimately, however, the teachings of Mahāvīra are not about words, or concepts. Mahāvīra was more concerned with the direct experience of freedom from fear and ego, which comes about through spiritual practices rather than through academic knowledge. Mahāvīra can show the sacred path on the map, but the pilgrim has to walk the path, unafraid of blisters.

I went through that direct experience of being a Jain pilgrim when I went on foot across the continents practicing and promoting non-violence, simplicity and the spiritual practice of walking, *tapas*. My friend E. P. Menon and I started from the grave of Mahatma Gandhi. We walked through Pakistan, Afghanistan, Iran and what was then the Soviet Union—Azerbaijan, Armenia, Georgia and Russia. During that journey we met two women who invited us for tea in their tea factory.

While we were drinking tea, one of the women went out of the room and came back with four packets of tea. She said, 'I would like you to carry these four packets of tea for me. Don't drink any tea from these packets yourself because they are not for you.'

'Who are they for?'

'Please deliver one packet of peace tea to our Premier in Moscow; the second packet of tea to the President of France; the third packet to the Prime Minister of England; the fourth packet to the President of the United States of America. And please give them a message from me because I can't get there myself. I am a very poor woman, and have children. But I have a message for the four great leaders of the world. Please deliver my message to them.'

'What is your message?'

'My message to them is: "If ever you get a mad thought of pressing the nuclear button, please stop for a moment and have a fresh cup of tea from these packets. This will give you a moment to reflect on the fact that the ordinary people of the world—men, women, children, farmers, workers—have done nothing to deserve your nuclear weapons. And your nuclear weapons will not only destroy the armies, they will destroy everything: people, plants, animals— they will destroy the Earth".'

This was a very powerful message of non-violence. When this Russian woman gave us peace tea, she was telling those leaders of the world that we have no right to damage the Earth, which we have inherited as a gift from the universe.

If we are given a gift, we treasure it. When we get a birthday present, we have an attachment to that present. We value it because it is a gift. A gift is given with love. The Earth is given to us as a gift, out of love for all beings. The Earth is a cosmic gift. I cannot think of any better way to express the idea of reverence for life and compassion for all beings, than the way in which that Russian woman expressed it.

Trusteeship

The idea of reverence is accepted even in science. In the scientist James Lovelock's new book Homage to Gaia, he says that while he is a scientist, and does not believe in God, he yet has a reverence for Gaia.[19] He reveres the Earth. The gift of the Earth is entrusted to us, to humans. Earth is a sacred trust. We are not the owners of the Earth, not even the guardians or stewards of the Earth. From a Jain perspective, we are the trustees of the Earth. Trusteeship is a responsibility assumed on behalf of all living beings and also for the future generations.

If you are a trustee of any trust, you will know that trustees are not allowed by law to use the trust money for their own purposes. A trustee is not allowed to benefit personally from the trust. The only thing that a trustee can do is to receive what is called 'expenses'. In other words, 'what you really need'. Otherwise trustees use the income of the trust for the benefit of the general public. A good trustee will not squander the original capital. Only the interest or the income can be spent. All of the great foundations, such as the Ford Foundation, have large amounts of money left to them as a gift. The trustees are given the responsibility of not depleting the capital.

In the modern world we use the environment, and the resources of the Earth, as if the Earth were income rather than capital. We use it, use it and use it until the Earth is depleted. If you are a trustee, then you do not deplete the capital. Rather you replenish the capital. Good trustees are those who receive a million pounds and distribute the grants to various bodies that are doing good work, and yet ensure that the original capital is enhanced. We need to do the same thing for the Earth. Because the Earth is our true capital. We need to consider ourselves as trustees of the Earth. Every one of us is a trustee with the responsibility to replenish the assets of the Earth. And yet the fruits of the Earth can be shared among all.

As with a tree, giving us apples, of course we can enjoy them. But we should keep the tree in good health by giving compost, good protection and good care; and when its fruits are ripe we can distribute them to those who need them. So the fruits of the Earth, the fruits of the forests, the fruits of the rivers, are given to us; and we may take them without harming the health of the Earth. We need to treat the Earth with respect, with reverence and with gratitude.

When we eat dinners we thank the cook, we thank the host, and we also need to thank the Earth which has produced this wonderful food: apples, tomatoes and aubergines; and rice, lentils and beans, so many good things. The cook will not be able to produce that food out of thin air unless the Earth has given that gift of food to us. Therefore we need to have a sense of gratitude to the Earth. The Earth is sacred. The gift is sacred. We are not to take it as if it was our right. We are not to take the Earth for granted.

The Jain ecology is a sacred ecology. It is not merely a pragmatic or utilitarian ecology. Utilitarian ecologists would say, 'We will protect the trees because trees give us oxygen, wood and furniture. Trees are useful to humans: therefore we will plant more trees'. That is a utilitarian approach. If there is no shortage of trees you will not care. But 2600 years ago Mahāvīra taught us to protect trees even when there is no shortage, just as you respect human life. You do not say, 'I will take a human life because there are too many humans': you respect human life and consider it sacred. We have to carry the same idea forward to animal life, water life, plant life, bird life, fish life—all life—life everywhere is sacred. And then receive our sustenance from life as a gift and with gratitude. We take wood from the tree to build a house, and we say, 'Thank you, tree, for sacrificing your body for our comfort; we are grateful to you'. When you have that sense of gratitude, then ecology becomes a sacred trust. Otherwise ecology is reduced to utilitarianism. In sacred ecology, Nature is good in itself. All life has intrinsic value. The value is not given to life because of its utility for humans. When we start to think in those ways then we see how dependent we are on the gifts of the Earth.

The American Founding Fathers made a 'Declaration of Independence'. That was a different context, the context of independence from colonial rule. When we are talking about ecology as a sacred trust, then we have to shed our arrogance of being independent. There is a kind of arrogance when we say 'I am independent', as if I do not need water, air or soil. If the worms were not working hard to produce food in the garden, we would have no vegetables for lunch. We depend on the worms for our survival. We may think that humans are superior in the evolutionary ladder, and the worms are down there under the Earth, they are of no significance. But human life will not exist if the worms are not working hard under the ground. Therefore Descartes's dictum 'I think, therefore I am', which is in a way the declaration of independence, needs to be revised. 'I think, I have my own mind, I have my own ideas, this mind gives me my existence.' Such an approach is based in individualism and dualism.

According to Indian tradition, we say *So 'ham*, 'You are, therefore I am'. Worms are, therefore I am; trees are, therefore I am; water, and rivers are, therefore I am. And not only Nature but our parents are, therefore I am. If my mother had not carried me in her womb for nine months, would I be standing here in front of you, giving a talk? No. I do not exist by myself, I exist because of my ancestors, because of my culture, because of my teachers, because of the Buddha, because of Mahāvīra, because of Jesus Christ, because of Shakespeare, because of Tolstoy, because of Gandhi, so many great teachers who have taught me, I carry their tradition with me. Jain ideas are based in mutuality, in relationship, and in reciprocity. We are part of the web of life. We have no independent existence.

Modern humanity has grown too arrogant, thinking that the Earth is there for our benefit. We are its masters, we are in charge, we can use the Earth as we like. This is the arrogance of humanism. From a Jain perspective, we humans need to learn humility: humility and humanity go together. Without humility, humans cannot be humans. In humility we depend on each other, we depend on the Earth. We cannot just take, take and take and destroy the Earth.

When we destroy the Earth, Earth will destroy us. We will get global warming, climate will change, floods will come, weather conditions will change. So we need to learn humility. And in that humility we say, 'We depend on the Earth'. We are trustees of the Earth.

BRIAN GOODWIN
Circling the Square: Moving from Control to Participation in Science and the Arts

✿

Setting the Scene

A group of people with a diversity of interests gathered at Mildura, Australia, in March 2000, to explore the confluence of art and science. One of the participants, an ecologist by the name of Ben Gawne, described the key qualities that a scientist must possess: a passion for ideas, a desire for meaning, an awe of the natural world, and a desire for control and certainty.[20] The last of these stands out as the quality that most obviously distinguishes a scientist from an artist. In what follows, I shall be considering when and why the desire for certainty and control became a defining characteristic of Western science; what effect this has had on the place of art in Western culture; and how current developments are changing the very nature of science so that the distinction between art and science is dissolving, allowing for a confluence towards a form of participation in creativity.

The image that I shall use for these developments is in my title: 'Circling the Square'. A problem that arose in ancient Greek and Egyptian culture was the converse of this: to square the circle, that is, to find a circle with the same area as a given square. An equally difficult problem, whose solution involved the discovery of the Golden Ratio, was to divide a rectangle into a square and a rectangle with the same ratio of sides as the original rectangle. These were very challenging questions for mathematicians whose sole means of construction were ruler and compass, as in Euclid's *Elements*. Now, however, with the remarkable flourishing of mathematics in Western culture, and particularly in its relation to the development of Western science, they can be easily solved by school children. As a culture we have done that, been there. It is time to move on and, as I shall show, science is itself pushing us into new and unfamiliar territory that requires different skills. This is the challenge of 'circling the square', which conjures up the image of the *passaggiata*, the evening movement around the square in a village, where the whole community comes together in conversation. To converse is to 'turn together', the many becoming one in a community of discourse and action. But the square is now larger since we have come to live in a global village. That village includes nature, not just humanity. So to circle the square now involves conversing not just with each other but with all the other inhabitants of our village. Just as we cannot control each other and remain creative, we cannot control nature and allow its creativity to flourish. What seems to be required is a move into participation, which involves cultivating the skills of the arts and sciences combined in a new adventure on which our lives depend. This is exciting as well as dangerous; but that's life, so let's try to see whither it is now beckoning.

Science: Out of Control, Into Participation

Western science is distinguished by its systematic pursuit of reliable knowledge of nature, which can then be used to control natural processes. The definition of 'reliable' in this context is interesting and illuminating: it refers to knowledge based on measurable quantities such as mass, velocity, force, length, direction, and so on. These quantities are organized into relationships based on mathematical principles which provide coherent descriptions of natural processes such as the movement of a pendulum, the planets, falling bodies, and projectiles. This science of quantities extends from physics to chemistry, biology, medicine, earth science and a host of other disciplines. The success of this approach to understanding and controlling the natural world is extraordinary, and has provided us with a remarkable range of technological products: from cars, computers and satellites to herbicides, drugs and genetically engineered organisms.

It is the nature of science to march on into unexpected territories, and one of these has emerged just as public confidence in the capacity of science to manage our lives is waning dramatically. Questions about the ability of scientific principles to provide answers to major problems have arisen in three areas: health, agriculture and ecological management. At the same time, new insights have emerged in science about the properties of natural systems which are described as complex—that is, having many diverse parts that interact to produce coherent wholes. These include organisms, communities and ecological systems, corresponding precisely to those areas where questions have arisen about the ability of science to control and manage.

The first indication that there are aspects of nature that are intrinsically out of our control came from mathematical insights that arose from studies of the planetary motion, where Western science began. When three planets interact, such as Saturn, Jupiter and Uranus, it is possible that instead of behaving predictably their motion can be chaotic. This possibility was first revealed in mathematical studies by Henri Poincaré towards the end of the last century, but it took nearly a hundred years for chaos to be clearly understood mathematically as a dynamic aspect of natural systems. Now this is commonplace knowledge. What it means is that not only the planets but any natural process can behave in such a way that we are unable to predict its behaviour beyond a limited period into the future. One example is the weather. We always knew that it is unpredictable, but now mathematics can tell us precisely why.

The second area of scientific study which reveals the unpredictability of natural processes is known as the science of complexity. Here we come to those complex systems such as ourselves (organisms), communities and ecological systems that I mentioned earlier as areas where applied science is failing to manage satisfactorily and crises are perceived to be arising in health, in the quality of life in communities, in agriculture and in the environment. The science of complexity can now give us insights into why this is so: these systems respond unpredictably to disturbances or to changes in their components. They have what are called 'emergent properties' which cannot be predicted from a knowledge of their parts and their interactions.[21] So the assumption that we can change our farming methods or use drugs or genetically modify organisms in predictable and controllable ways is unfounded. Here science meets a natural boundary of prediction and control. However, we cannot live without interacting with these complex systems, as our lives depend upon them. If we cannot control them, what is the appropriate behaviour? An alternative is to participate with them; to enter into a conversation with them. Since a desire for control and certainty seems no longer to be an appropriate quality for a scientist, we can go back to the other characteristics defined by Ben Gawne: a passion for ideas, a desire to participate in natural creativity. This list is probably acceptable to most artists as well. A first step towards convergence has occurred. But it is necessary now to look a bit more closely at what it might mean to participate and converse with our fellow beings (all of them) in ways which combine science and art.

A Science of Qualities

Since qualities were left out of Western science, and are increasingly absent from our lives, this seems like a good place to start in looking for a way to correct an imbalance that is now threatening the quality of life on the planet. First, what do we mean by qualities? They are properties associated with experiences such as pleasure, pain, joy, sorrow, beauty, taste, smell, and so on. These cannot be characterized by measurement, by quantities, although they have different degrees of intensity. In our culture they tend to be associated with what we call subjective experience, personal and idiosyncratic, distinctive to individuals and therefore not 'objective' or reliable. Our science is founded on the proposition that reliable knowledge of the world comes only from quantifiable experience, as described above. However, the paradox is that qualities are every bit as important in our lives as quantities, probably more so. Our relationships with each other are based on the quality of our experience in interactions, which allow us to judge whether others are friendly, trustworthy, honest, open, or not. Quality of experience is basic to our judgement of quality of food and drink, of how we spend our leisure time, of the jobs we have, of the company we keep (whether human or non-human), of the music we listen to, of where we go for holidays, and so on.

Why is this extremely important area of life excluded from what we call the natural sciences, such as physics, chemistry and biology? Why is it that when we study colour we concentrate on the frequency of the electromagnetic waves associated with particular colours, the velocity of light in different media and the differential absorption of light waves by materials of different composition, such as pigments? Why not pay equal attention to the feelings, the experiences we have when we see colours of particular hues? The properties to which science pays attention can all be quantified. The decision to focus on quantities was taken in the sixteenth century, when Western science was beginning to be defined in a precise and distinctive way. It was an historical choice, not something required by nature itself. We can re-examine that choice and change science if we wish. I believe that the current situation has gone beyond a wish: it has become a natural next step in the development of Western science, and an urgent need. In fact, this extension of science to include qualities as well as quantities is already happening on a variety of fronts.

There are traditions which have always included qualities. These include all of the healing traditions, conventional as well as complementary. Quality of pain cannot be ignored in any diagnosis, and it is widely acknowledged that healing practice is as much an art as it is a science. Western medicine tries, naturally enough, to replace qualities by quantities, but this simply isn't possible and there is now a move to extend the evaluation of any therapy to an assessment of quality of experience as reported by the patient. The therapeutic relationship is becoming generally recognized as a cooperative journey and a conversation between the therapist and the subject wherein qualitative experience is of equal importance with quantitative assessment of, say, blood composition and other measurements of body state.

There is still some way to go, however, before scientists accept qualitative evaluations as reliable, 'objective' indicators of what is perceived and experienced in nature. This requires a demonstration that many different people observing the same thing, such as a pig in a pen interacting with somebody or a particular colour, agree about the quality of experience of the pig (nervous, boisterous, laid-back, etc.), or about their own experience (e.g., blue = peaceful, relaxing or something qualitatively similar). Consensus is the way in which scientists come to agreement about what constitutes 'objective' knowledge. You and I are unlikely to agree on the precise weight of an object by simply picking it up and giving an estimate, but we have a reliable method of coming to an agreement by using scales. There are methods whereby the quality of food and drink is evaluated by 'tasters' using consensual procedures. The 'scales' in this case are internal, but nevertheless reliable in that there is a consistency of consensus between different 'tasters' on different occasions. These procedures are now being extended to other forms of qualitative experience.[22] A science of qualities is on its way, and not before time!

Cultivating the Intuition

The extension of Western science to include qualities was actually started over two hundred years ago. Johann Wolfgang von Goethe explored ways of systematically studying nature that use the full range of experience to arrive at an understanding of the wholeness of natural processes. Here is how he describes this adventure:

> None of the human faculties should be excluded from scientific activity. The depths of intuition, a sure awareness of the present, mathematical profundity, physical exactitude, the heights of creative reason and sharpness of understanding, together with a versatile, ardent imagination and a loving delight in the world of the senses—they are all essential for a lively and productive apprehension of the moment.[23]

One of Goethe's discoveries in the study of light was that there is a natural circle of colours such that they all relate to each other in a holistic manner, rather than the linear arrangement of colours from red to violet that we learn from the Newtonian colour spectrum. It was his interest in the use of colour in painting that led him to this work, for he wanted to discover the regularities that underlie the painter's use of colour to express feelings and mood as well as to depict natural form. The painter's work is intuitive, based on a reliable understanding of colour relationships and of the feelings they evoke. These feelings are not subjective add-ons. They are regular components of colour experience. This becomes clear as we return to the domain of healing. Colour therapy uses different colours to evoke different feelings, which are associated with different physiological states that need to be enhanced to balance the state of the subject receiving therapy. While every individual has distinctive emphases and associations in the experience of a particular colour, there is also a common component which is shared by all. Qualities reflect real aspects of the experienced world.

There is another very significant property of qualities: they often refer to the characteristic of the whole that is being perceived. When we say that a dog is playful, we are expressing a condition of the whole dog—its behaviour and its feelings, how it is seen to live in the world of relationships. We say that we intuit this state of the dog, or that we see it directly. We don't count the number of times it barks per minute, how many movements it makes, etc. We 'see' the dog's experience through its actions as a gestalt perception, seeing the whole. Similarly, when we talk of health we refer to the state of the whole person, or of an organization, or of a landscape. Quantities are very good for characterizing aspects of parts, such as the rate of beating of the heart, while qualities such as feeling well, happy, sad or stressed express the condition of the whole person. We instinctively use such descriptions not just for ourselves but for other animals and, by extension, for places, buildings and organizations.

Western science, however, sees these as purely subjective and so discounts them as mere personal associations or as projections from the person onto a place or a building or a housing estate. The result is that we have become progressively anaesthetized to our surroundings, accepting ugliness and destruction in the built and

natural environment as inevitable correlates of progress and technological development. We need to re-awaken our sensitivities to
recover quality of life. This involves cultivating our intuitive perception of the qualities of our own lives, and of those of all the other
beings who share the planet. Here is a major job that brings the arts
and the sciences together in a new enterprise that largely abolishes
the distinction between them. Scientists need to cultivate intuition
systematically as part of a science of qualities, while artists need to
stimulate a re-awakening of sensitivity to the worlds of human
artefact and to nature. Both would then be participating in a celebration of the creativity of the cosmos, engaging in a conversation
with the world in which every participant, from rock and tree to
human and planet, has an equal voice. We have tried the path of
control in which scientists discover laws that provide the power of
prediction; and we have found that, despite its extraordinary
insights and its capacity to provide us with a great range of technologies, it is very limited in its application to a reality that is
creatively unpredictable. Artists have explored inner worlds where
they are masters because they are the arbiters of individual creation;
and we have found that this detaches us from, rather than bringing
us into balance with, the creativity of nature.

Living for Now, Not the Future

It seems that we need a new enlightenment, and a new enlightenment requires a new education. In this the arts and sciences will
again be joined, as they were before the Renaissance separated them
along the cleavage lines of thought and feeling: the former giving us
a rational, mechanical, predictable world; the latter subjective, idiosyncratic, unpredictable novelty of individual expression. Both of
these apparently opposite movements are orientated towards the
future. Science constantly promises to explain what is currently
unknown, and its technological applications promise a better
future for humanity. The arts promise unending novelty and enter-
tainment. The preoccupation is with a future which never arrives,
anaesthetizing us from the reality of the present.

Sooner or later, however, we had to wake up from this dream of a 'messianic age' (for it is a dream with a Judaeo-Christian origin, however heretical it has become in its materialist form), and see the mess that has resulted from our neglect. It is now very visible to all, but we are still caught in dreams of the future: new technologies that will allow us to go on living in the same way but without the mess; new art that will explore and reveal ever more intimate aspects of ourselves and of our relationships with each other. The paradox of creativity is that, while it does produce novelty, it has another dimension that is more significant: it needs to be appropriate to context, to now. This is where we need to cultivate our sensitivities, to feel our way from here to a healthy, whole, healed future by a path that is at present invisible but is revealed as we walk it. This is the path of what is sometimes called 'right action'. To get to a future in which things are better, the only reliable way to go is by fully tuning in to the present so that the future arrives as an unexpected reve-lation from engaged action in the now, not from prediction and planning. This is going to be quite a difficult lesson for our future-addicted culture. But it is the lesson that is coming to us from all quarters: from developments in complexity and creativity in science; from movements in the arts to reconnect with total context; and from action in civil society in response to the perceived mess of our farms, our countryside, our cities, our environment and the planet. An integrated educational system that puts all these components together in learning for participatory living is already on its way, as an appropriate creative response to the present situation. We just need to stay in touch with this, connected and responsive, and not get lost in another invented future.

SUHEIL BUSHRUI
Environmental Ethics: A Bahá'í Perspective

❧

World unity, with its corollary world peace, is the central challenge of the Bahá'í religion, a task that is intimately connected to environmental ethics. The oneness of creation, the reciprocity of the material and spiritual elements of society, and the unity of mankind are interrelated aspects of this principle. A Bahá'í environmental ethic is implicit in these three concepts. They will be vital in the ramification of this ethic in the future, as efforts are made to understand them better and to apply them to environmental problems and to humanity's continuing relationship with the environment.

A consciousness of the oneness of creation and of the mutuality of the material and spiritual elements of society, and the counterpart of such consciousness in action—its beneficent expression in our wise use and care of the environment—are dependent on humanity's unity. It is therefore appropriate, in addressing the Bahá'í approach to the environment, to reflect for a moment on the beauty of human unity in the context of the Bahá'í Faith.[24] Using metaphors of the unity of nature, 'Abdu'l-Bahá, the eldest son and lawful successor of the prophet-founder of the Bahá'í Faith—'the servant of Bahá'u'lláh', as he chose to be known[25]—conveys the essence of this unity:

> Man must strive that his reality may manifest virtues and perfections, the light whereof may shine upon everyone. The light of the sun shineth upon all the world and the merciful showers of Divine Providence fall upon all peoples. The vivifying breeze reviveth every living creature and all beings endued with life obtain their share and portion at His heavenly board. In like manner, the affections and loving kindness of the servants

of the One True God must be bountifully and universally extended to all mankind. Regarding this, restrictions and limitations are in no wise permitted.[26]

The other bases of a Bahá'í environmental ethic, no matter how well understood and practiced, will not be sufficient without the establishment of a 'Commonwealth of all the nations of the world' implicit in the Bahá'í concept of unity.[27] For the Bahá'í believer, only this unification of mankind will permit the practical expression of the universal love of humanity as a whole which the above statement of 'Abdu'l-Bahá calls for.

The physical order is under attack from unbalanced commercial exploitation, pollution, pesticides, and other effects of immoderate and ill-motivated human interference in the natural cycle of the environment. The harmful consequences of this vast destruction are incalculable and varied. For example, the producers of food for the developed world find themselves working in harsh conditions with severe health risks and for insufficient wages. The discrepancy between those who have and those who have not can be seen in the plight of the people of Honduras, currently attracting widespread attention because of the devastation of their country by hurricane, flooding and epidemics. Its economy depends largely on bananas, which supply Honduras with 70% of its foreign exchange. Bananas are a productive crop (a sapling bears fruit nine months after planting), and a valuable source of nutrients, especially potassium. Yet this crop is grown mainly for export, denying the local producers and their families the sustenance which they need. If a banana sells for 10p in a British shop, 3p of that would go to the retailer and the same amount to the packagers. Wholesaler, importer and shipping company would each receive 1p, leaving 1p to be shared between the grower and picker: 5% for each out of the total sum.[28]

For the Bahá'í believer, neither science alone nor a respect for nature can address these problems. Only a 'world federal system' animated by concern for all the people of the world will enable mankind to arrange its economic, material and social life in a manner concomitant with justice for all peoples and the duty of reverence towards the earth and all other elements of the physical order.[29] Before considering this basis of a Bahá'í environmental ethic, however, it is fitting to give some thought to the requisite attitudes toward the physical world. This question will be approached from two angles: first, by reflecting on the oneness of creation; and second, by considering an example in architecture and building of the mutuality of the material and spiritual elements of a society.

The Oneness of Creation and Humanity's Unity with Creation

The Bahá'í religion shares a basic belief in the oneness of creation with the world's other religions and spiritual traditions. Humanity has been educated concerning the oneness of creation through the many perspectives on it provided in scriptures, traditions and poetry associated with these belief systems. Four of these perspectives stand out as being among the most salient: the interconnection between humanity and the rest of creation; the sanctity of all creation; the duty of respect for the material creation; and, last but not least, the use of the material creation as a medium for humanity's spiritual education and expression.

The relationship between man and creation is depicted in the Book of Genesis, for example, where we read that God gave man dominion over the created world,[30] a statement which has been hotly disputed by Biblical scholars. What is generally agreed now is that 'dominion' implies a responsible stewardship rather than a thoughtless exploitation, although the Biblical passage has been subject to widespread (mis)interpretation to justify commercial practices, even slavery. The bare story in Genesis, however, also indicates humanity's oneness with the rest of creation—both were created by one Creator, and hence they are connected with one another.

This idea is further developed in the writings of 'Abdu'l-Bahá:

> Know ye that the world of existence is a single world, although its stations are various and distinct. For example, the mineral life occupieth its own plane, but a mineral entity is without

any awareness at all of the vegetable kingdom, and indeed, with its inner tongue denieth that there is any such kingdom. In the same way, a vegetable entity knoweth nothing of the animal world, remaining completely heedless and ignorant thereof, for the stage of the animal is higher than that of the vegetable, and the vegetable is veiled from the animal world and inwardly denieth the existence of that world—all this while animal, vegetable and mineral dwell together in the one world . . . . . Of this power of discovery which belongeth to the human mind, this power which can grasp abstract and universal ideas, the animal remaineth totally ignorant, and indeed denieth its existence.[31]

Thus the structure of the world is revealed as a series of precise gradations, with the lower degrees unaware of the higher levels of existence and their powers of perception. By extension, man himself is unaware of insights that are reserved solely for the Creator and cannot, despite his superior intellect and powers of reasoning, unravel the ultimate secrets of the universe. He therefore shares the characteristics of the created that distinguish the rest of creation from its Originator.

This insight should help to maintain in us a proper spirit of humility before God: limitless as our faculties may seem to us, they are negligible in relation to the omnipotence of the Creator. Bearing in mind the connection between all parts of the creation, man is enjoined not to adopt an arrogant and overbearing attitude towards lesser species.

The Oneness of Creation and the Sanctity of Its Every Part

Far otherwise, man is to see all creation as good. Just as humanity and all beings are one in their distinction from their Creator, so are they linked with the Divine in the fact of their creation by the Divine. From this may be derived a second perspective on the oneness of creation: the sanctity of all creation. This teaching is developed in a variety of ways in the traditions of most religions. We read, for example, in the Bṛhadāraṇyaka Upaniṣad: 'It is not for the love of creatures that creatures are dear; but for the love of the Soul in creatures that creatures are dear.'[32]

Humanity and all other creatures are here seen as united in their reflection of divine attributes. The writings of Bahá'u'lláh, the founder of the Bahá'í religion, give us further insights into this teaching of the sanctity of all creation:

> Say: Nature in its essence is the embodiment of My Name, the Maker, the Creator. Its manifestations are diversified by varying causes, and in this diversity there are signs for men of discernment. Nature is God's Will and is its expression in and through the contingent world.[33]

> From the exalted source, and out of the essence of His favor and bounty He hath entrusted every created thing with a sign of His knowledge, so that none of His creatures may be deprived of its share in expressing, each according to its capacity and rank, this knowledge. This sign is the mirror of His beauty in the world of creation.[34]

And in another verse: 'No thing have I perceived, except that I perceived God within it, God before it, or God after it.'[35]

To make such an act of affirmation, recognizing the presence of a divine Creator throughout the entire universe, commits the Bahá'í believer to becoming an involved partner in the process of nurturing and protecting that creation, which, even in its most minute particles, reflects the nature of that Creator:

> Every man of discernment, while walking upon the earth, feeleth indeed abashed, inasmuch as he is fully aware that the thing which is the source of his prosperity, his wealth, his might, his exaltation, his advancement and power is, as ordained by God, the very earth which is trodden beneath the feet of all men. There can be no doubt that whoever is cognizant of this truth, is cleansed and sanctified from all pride, arrogance, and vainglory.[36]

Consciousness of the sanctity of the natural world creates in the believer a further awareness of his oneness with all creation. This awareness develops in man the humility intrinsic to this oneness: 'Be forbearing one with another and set not your affections on things below. Pride not yourselves in your glory, and be not ashamed of abasement. By My beauty! I have created all things from dust, and to dust will I return them again.'[37]

Love of nature in the Bahá'í sense, then, is inseparable from the love of God. It is not a partial or sentimental attachment to randomly chosen aspects of creation, but an affirmation and wholehearted acceptance of creation as a whole: 'Out of the wastes of nothingness, with the clay of My command I made thee to appear, and have ordained for thy training every atom in existence and the essence of all created things.'[38]

The believer is constantly aware of the supremacy of the Creator, and that divine inexhaustible power to create life out of nothingness. All of the elements of existence have a purpose and are designed for our 'training'. These gifts place on every human the obligation to treat them with respect, mindful of our place in creation and of our responsibility towards our fellow creatures.

The Oneness of Creation and the Duty of Respect

Responsible husbandry of the earth and its resources is recognized as an essential precept in many of the world's religious traditions. The earth itself was often worshipped as a deity, and the thought of drawing ceaselessly and greedily from its store without giving something back in return was beyond contemplation. We find this motif in the Greek worship of Demeter and Persephone, the harvest goddess and her daughter, who represented spring and renewed fertility. Although the ritual practiced at Athens, where women cast pigs as sacrificial offerings into a deep pit in the temple precincts,[39] may seem barbarous and distasteful to us, it nevertheless illustrates the principle of restoration and repayment. The underlying belief was that the earth, if not treated with due reverence and gratitude, might turn harshly on those who failed to give her proper respect.

With the emergence of the Gaia movement in recent years, the belief has once again become widely current that the earth, if mistreated, is capable of terrible retaliation. Countries such as Bangladesh demonstrate the viability of the Gaia philosophy. In that country, dams and road-building have affected the delicate balance between land and sea; and new wells drilled to provide pure drinking water have been found contaminated with arsenic present in the subsoil, slowly poisoning fifty million people. In 1998 alone, floods affected thirty million people, and the actions formerly taken to curb the flooding have actually exacerbated conditions.

Against the growing recognition of dangers inherent in upsetting the equilibrium between man and nature, what the sacred writings of the world's religions have to say concerning humanity's duty of respect for creation merits reflection. Thus there is the following text from the Qur'án: 'Do not spread corruption on earth after it has been so well ordered. And call unto Him with fear and longing: verily, God's grace is ever near unto the doers of good!'[40]

This verse draws our attention both to the first perspective on the oneness of creation: the connection between humanity and all creation; and to the second perspective: that the universe and humanity share with each other a dependence on God and are therefore both implicitly sacred. It also hints at a third basis for the oneness of creation: respect. In drawing a parallel between the proscription against 'doing mischief to the earth' and the duty to seek one's order and good from the Creator, with faith that 'doing good' is sufficient as a means to have needs met, this text indicates that man's relationship to the earth is subject to moral rules and as such it suggests our duty to respect the earth.

The same concerns are reflected in a Zoroastrian confession text:

For the sins which I have committed against Ohrmazd the Lord, and against men and all manner of men . . . against cattle and all manner of cattle . . . against fire and all manner of fires . . . against metal and all manner of metals . . . against earth and all manner of earth . . . against water and all manner of water . . .

and against plants and all manner of plants, I repent, am contrite and do penance.[41]

Among the Maori of New Zealand, a similar view is enshrined in three ancient proverbs:

The treasure of land will persist; human possessions will not.

The treasured possessions of men are intangible; the treasures of the land are tangible.

Without language, self esteem and land people will cease to exist.[42]

And in the writings of 'Abdu'l-Bahá, we read a summation of the Bahá'í precepts on this matter:

Briefly, it is not only their fellow human beings that the beloved of God must treat with mercy and compassion, rather must they show forth the utmost loving-kindness to every living creature . . .. The feelings are one and the same, whether ye inflict pain on man or on beast.[43]

Here we find a view of unity which is central to the Bahá'í teachings: to wound a fellow creature, whether human or non-human, strikes at the whole order of creation, which in a way is a transgression against the Creator Himself.

The Oneness of Creation and the Language of Humanity

A fourth perspective on the oneness of creation implicit in diverse religious writings and spiritual traditions emerges from their reliance on nature imagery to convey spiritual and intellectual meaning. In this way, nature serves as a medium of spiritual instruction and expression.

Among the countless environmental analogies in the writings of Bahá'u'lláh, nature is not only a paradigm of divine creativity and generosity but a source of metaphors and similes to illustrate aspects of spirituality and divinity, love, virtue, and the Creator's relationship with His creation:

The Word of God may be likened unto a sapling, whose roots have been implanted in the hearts of men. It is incumbent upon you to foster its growth through the living waters of wisdom, of sanctified and holy words, so that its root may become firmly fixed and its branches may spread out as high as the heavens and beyond.[44]

Perception of divinity is not confined to the absorption of the written or spoken word through sight or hearing. Indeed, it may be achieved through other senses as well:

God grant that, in these days of heavenly delight, ye may not deprive yourselves of the sweet savors of the All-Glorious God, and may partake, in this spiritual Springtime, of the outpourings of His grace.[45]

And again:

O my brother! A divine Mine only can yield the gems of divine knowledge, and the fragrance of the mystic Flower can be inhaled only in the ideal Garden, and the lilies of ancient wisdom can blossom nowhere except in the city of a stainless heart.[46]

The sea is an especially frequent symbol:

And when I behold the sea, I find that it speaketh to me of Thy majesty, and of the potency of Thy might, and of Thy sovereignty and Thy grandeur.[47]

Fresh water is also used many times as a metaphor for God's reviving power and man's response to it:

We have caused the rivers of Divine utterance to proceed out of Our throne, that the tender herbs of wisdom and understanding may spring forth from the soil of your hearts. Will ye not be thankful?[48]

This is the Day whereon the rushing waters of everlasting life have gushed out of the Will of the All-Merciful. Haste ye, with your hearts and souls, and quaff your fill, O Concourse of the realms above![49]

Throughout the writings of Bahá'u'lláh, such nature imagery is used to refer to the Creative Word again and again. Trees, birds, flowers (especially roses), water and the ocean are used not only as individual symbols, but as a symbol of the Faith of God: the Creative Word itself. Each has its own web of associations: birds suggest the soul, and water the force of life and source of all creation, recalling the statement in the Qur'án: 'Out of water We have created every living thing'.[50] The tree, too, 'a tree neither of the east nor of the west',[51] is an enduring symbol not only in the writings of Bahá'u'lláh but in many other cultures: the Judaeo-Christian tree of the knowledge of good and evil; the source of the golden apples of the Hesperides or the fruit which supplied the Norse gods of Asgard with eternal youth; the great world ash tree Yggdrasil which, in Scandinavian myth, linked the underworld with Midgard, the world of mortals, and Asgard, the realm of the gods. The green and golden tree of life evoked by Goethe in his *Faust* is a potent metaphor for the vigor and enduring quality of God's creation.[52] In a Bahá'í context, though, this image gains an added dimension from the repeated emphasis on the divine intent and loving purpose underlying creation. As the lethal effects of acid rain become ever more apparent, the stripping of foliage and the death of trees in the great forests of Scandinavia and Central Europe are a poignant reminder of our failure to be conscious of this purpose. It is difficult to read the passage just cited from the Qur'án without reflecting sadly on the loss of the crystal clarity of the waters of rivers and lakes in many countries. This loss is a deadly reminder of how the destruction of plant and animal life is taking place.

To exploit nature for one's own ends is in Bahá'í eyes, as noted above, an assault not only on creation but also on the Creator. As such, it is a kind of blasphemy, which not only defaces the environment and degrades the perpetrator but directly affronts the Source of a bounty intended for the protection and prosperity of humankind.

Furthermore, this blasphemy vitiates the capacity of the religious writings of the past and the present to educate—to convey spiritual understanding. There is little hope for meaningful development when these vehicles of expression cease to hold a standard meaning. We cannot derive inspiration from these texts when, for example, the only air we know is poison, the only lakes we see inflict disease, the rains and snows bring nuclear waste, and the sun itself is a danger.

The Reciprocity of the Material and Spiritual Elements of Society:
A Holistic Approach in Architecture

A society's relationship with the environment manifests its values. Its ways of interfering with nature, whether in the organization of resources or in the development and refinement of sciences and arts, vary from age to age and region to region in accordance with the dominant culture, but never in known history has a society had the power over the physical world that the nations of the globe possess today. In an age when, as George Soros expresses it, 'market values have penetrated all departments of human endeavors', it is a challenge for every believer to become deepened in the oneness of creation, and to put into practice in everyday life this value, as well as other Bahá'í values relevant to the environment. Rejection of mindless consumerism, and of the thoughtless unbalancing of environmental systems that it entails, is an essential part of this process, both because of the strain that such consumption places on the world's resources and as a step towards the fairer distribution of those resources among its inhabitants. So, too, is the discarding of such practices as open-cast mining and the rearing of calves in veal crates. But it is to the new standards set by their Faith concerning the complementarity of the material and spiritual elements of society that Bahá'í believers look with greatest hope. These standards provide a means for relating the respect for all creation to other principles.

For example, the physical, environmental and economic activities of the society are to be harmonized with spiritual principles, which themselves need to be reasonable from the standpoint of our material knowledge. When these material activities are carried out in accord with spiritual principles, the goals of a proper human relationship with the environment will be met.

One particular sphere of activity that has an immense impact on the environment is building. This activity is carried out by all kinds of creatures: ants build anthills; bees, highly organized colonies of hexagonal cells; weaver birds, complex nests; rabbits and mole-rats, networks of warrens and burrows; and, of course, castles and cathedrals, temples and towers, belong to the world of man. As human building becomes increasingly sophisticated, its contrast with building in the natural world is striking. In nature and so-called 'primitive' societies, whether animal or human, building materials are generally drawn from the immediate environment (mud, reeds, bricks of alluvial clay, local stone). But twentieth-century man has at his disposal a wide range of materials, both organic and inorganic, gathered from all over the world, often at a considerable cost in energy and labor. His activities in this domain therefore have a greater impact on the environment and the society than they did in the past.

Although there is a consciousness and awareness of the effects of man's activities on his surroundings, Bahá'ís realize that the development of architecture, technology and the engineering sciences which accord fully with Bahá'í principles is a long way down the road. Nevertheless, the Bahá'í writings often touch upon the practical application of the principle of the complementarity of the religious and material elements of a society, and therefore help to focus the activities of believers in this direction. 'Abdu'l-Bahá writes about schools, for example: 'The school must be located in a place where the air is delicate and pure'.[53]

Here a spiritual text conveys a directive concerning the physical environment. Its counterpart must be the material knowledge and material conditions necessary to accomplish it. Only when we have the knowledge and will to create and maintain an environment with this air can we put this teaching into practice. On the other hand, the complement of putting in practice this spiritual guidance will be the protection of the most vital element of the environment—air.

In addition to such writings as these, the Bahá'í holy places can provide insight on how Bahá'í believers will harmonize their organization of the physical environment with spiritual principles. We have already commented on the need for a deep reverence for life in all its forms, a principle which underlies the Bahá'í Faith. Alongside this regard for the sanctity of life and creation, Bahá'í belief specifically enshrines a recognition of the importance of a spiritual frame of mind in every area of human activity. This inner, individual holiness is closely bound up with the concept of the physical sanctity of sacred spaces. Indeed, man in himself is such a space: as Christian terminology expresses it, the 'temple of the Holy Spirit'. The Báb, the prophet-herald of the Bahá'í Faith, describes 'this physical frame' as 'the throne of the inner temple', noting that 'whatever occurs to the former is felt by the latter'.[54] Shoghi Effendi, the 'Guardian' of the Bahá'í Faith, uses a similar concept in describing the body as the temple of the soul. Man in this world is a sacred temple, and the concept of building is closely linked with this idea.

By extension, this leads to an awareness of the possibility of creating a 'City of God', as St Augustine has it, which is not of this world but which requires the same high degree of discipline and organization as a concrete structure on earth. Reading the description of the New Jerusalem in the Revelation of St John the Divine, one is struck by the precise and detailed account of its dimensions and appearance, extending even to the individual jewels laid down at its foundations and adorning its gates. These have an allegorical significance, but also stand as symbols of the supreme richness and beauty of the city and the harmony of its parts. We find a similar concept in al Farabi's *al Madinah al Fadelah*, the Ideal City. The actual site of this city is frequently significant (Mount Carmel being a case in point), and along with their intrinsic sacredness, the sacred spaces are intimately associated with the dimension of human sanctity.

At this point, it is appropriate to consider, within the context of Bahá'í belief, the special importance of Mount Carmel, the site of the Bahá'í World Centre, and its associations with the founder of that religion, Bahá'u'lláh. We do not know of the environmental or physical import of this site, but the history of the Bahá'í Faith nevertheless explains why it is now the location of one of the Bahá'í holy places. After being exiled from his native country of Iran, Bahá'u'lláh settled in Baghdad, then in Constantinople (Istanbul), Adrianople (Edirne), and finally in 'Akká, which was then a penal colony of the Ottoman Empire. In 1890, on a visit to Haifa, he designated to his son, 'Abdu'l-Bahá, the place in which the remains of the Báb should be buried, and enjoined him to construct a fitting tomb. The Báb had been martyred in 1850 in Iran, six years after proclaiming his mission. To ensure that his remains did not fall into enemy hands, Bahá'ís had repeatedly to exhume and reinter the remains of the Báb in a succession of burial places until 1909, when they were at last transferred to the final resting place designated by Bahá'u'lláh on Mount Carmel. The Shrine of the Báb was built to the specifications of 'Abdu'l-Bahá himself, with subsequent embellishments added by Shoghi Effendi, Guardian of the Bahá'í religion from 1921 to 1957, in accordance with the directions of 'Abdu'l-Bahá.

The architecture of the Shrine, the work of the Canadian architect William Sutherland Maxwell, combines Western and Eastern elements in a visual symbol of the Bahá'í respect for the values and virtues of widely differing traditions. Maxwell used granite columns derived from Roman architecture, Corinthian capitals from ancient Greece, and the pointed arches and sinuous arabesques of Oriental art. In 1987, work was commenced on the building of eighteen monumental terraces encircling the Shrine, nine below it and nine above, designed by Fariburz Sahba who is renowned for his Bahá'í House of Worship in the shape of a lotus flower in India. Conceived as a series of nine concentric rings embracing the Shrine, the terraces on Mount Carmel are laden with plants and shrubs which blossom throughout the year, ranging from a formal landscape in the centre to a semi-wilderness that blends into the surrounding forests. Water and light are also integral to the concept of the gardens. Emphasizing the conservation of environmental and natural resources, an advanced technological system irrigates the terraces with the most economical use of water. This still point in the busy centre of Haifa attracts wildlife of many kinds, offering a refuge and haven for both man and the world of nature and, in the process, illustrating the Bahá'í attitude of respect and reverence for nature (and culture) in its manifold forms.

Besides the Shrine of the Báb, Mount Carmel holds the Seat of the Universal House of Justice, the highest international governing body of the Bahá'í religion, as well as the International Bahá'í Archives, the Library, and other administrative institutional buildings, all designed by the architect Husayn Amanat in the neo-classical style. An open letter issued on 31 August 1987 by the Universal House of Justice outlining plans for the erection of the Library, International Teaching Centre, and Centre for the Study of the Sacred Texts, together with the construction of the terraces and an extension to the International Archives, recalls the event which established the holiness of Mount Carmel as a sacred space: 'Nigh on one hundred years ago, Bahá'u'lláh walked on God's Holy Mountain and revealed the Tablet of Carmel, the Charter of the World Centre of His Faith, calling into being the metropolis of the Kingdom of God on Earth.'[55] In a letter dated 21 March 1952 addressed to the Bahá'ís in the East, Shoghi Effendi foresaw the future glory and sanctity of the Shrine and its surroundings:

This beautiful and majestic path . . . will subsequently be converted, as foreshadowed by the Centre of the Covenant, into the Highway of the Kings and Rulers of the World.

These mighty embodiments of kingly power, humble pilgrims to the Sanctuary of the Lord, will, upon their arrival in the Holy Land, first proceed to the Plain of 'Akká, there to visit and circumambulate the Qiblih of the people of Bahá, the Point around which circle in adoration the Concourse on

High. They will then make their way to this august and vener-
ated city, and climb the slopes of Mount Carmel . . . .. Reaching
the threshold of the Sanctuary of Grandeur, they will cast their
crowns upon the ground, prostrate themselves to kiss its
fragrant earth and, circling around its hallowed arcade, call
out 'Here am I, Here am I, O Thou Who art the Exalted, the
Most Exalted One!', and recite in tones of fervent supplication
the perspicuous Verses of the Tablet of Carmel.[56]

Shoghi Effendi envisaged the pilgrimage of the kings to com-
memorate the sufferings and martyrdom of the Báb, but also to
witness the beauty of the 'garden of the Exalted Paradise', the glory of
Carmel and the power and greatness of the Lord of Carmel Himself.
Implicit in this statement is an awareness of the universal
significance of Mount Carmel as a holy site for members of other
religions: the twin cities of Haifa and 'Akká were sacred to Christians,
Jews and Muslims, as well as to Bahá'ís. Equally important is the
natural beauty of the spot, where visitors might 'inhale its fragrant
scents and taste of its choice and luscious fruits',[57] not as a mere
source of sensual gratification but as an inspiration for reflection on
the might and bounty of the Creator and the sacredness of all life:
'Every created thing in the whole universe is but a door leading into
His knowledge, a sign of His sovereignty, a revelation of His names, a
symbol of His majesty, a token of His power, a means of admittance
into His straight Path.'[58]

This consideration of Mount Carmel provides an example of
man's shaping of the physical environment in accord with a religious teaching, and shows us a faint glimmer of the great extent to
which the material and spiritual elements of a society can complement each other, to the mutual benefit of each and with favorable
consequences for the environment.

Environmental Ethics and the Unity of Mankind: The Example of Architecture

The verse from Bahá'u'lláh's writings just cited speaks not only to the inspiration which humans receive when experiencing the beauties of nature, but also to the material knowledge which they obtain when concentrating the intellect on the diverse facets of their physical environment, and the spiritual knowledge that is reflected in this material knowledge. We can, through the power of the intellect and its focus on the physical world, whether in arts or sciences, attain an amazing harmonization of the spiritual and physical elements of society. But for the Bahá'í believer this principle cannot take practical effect without giving equal attention to religion, and in particular to such a vital religious principle as the unity of mankind. Building activities need to serve the development and protection of all individuals on the planet and the world community as a whole. Correspondingly, they cannot be perfected without all humanity's receiving the bounties of universal justice, education and love. The pure air that should surround all the schools, and the fruits, the flowers, the water and ornaments that embellish a sacred building, will require more than a harmonization of science with spiritual knowledge; all this will need a change in the attitude towards humanity.

In this context, it will be useful to refer to the writings of Leo R. Zrudlo, who calls upon us to consider a challenge for the twenty-first century in his article 'The Missing Dimension in the Built Environment'.[59] From a Bahá'í viewpoint, he compares definitions of architecture offered by various authors and architects such as Frank Lloyd Wright, Le Corbusier, and Leon Krier, and demonstrates the growth of concern for a spiritual dimension which has been absent from urban planning for several decades. If, as Goethe puts it, architecture is frozen music, without that spiritual content it becomes music in disharmony and a source of further discord and malaise among those who have to live and work in such an environment. The much-discussed 'sick building syndrome' is an example of this, where disregard for the spatial and spiritual needs of work actually produces physical symptoms of ill health.

Zrudlo's thesis is based on two important principles: unity in diversity, and consultation. With the first, we once again find ourselves contemplating the tenet of unity and interconnectedness: neglect one aspect of creation or the creative diversity of life, and the whole will suffer. The second principle is inextricably associated with the first: consultation implies a respect for the needs and values of those for whom the building is intended or who will be affected by its construction. This second principle thus applies the principle of unity in diversity as it is most prominently in the Bahá'í writings—to the diversity in the human family.

Zrudlo suggests ways in which planners and architects can incorporate a spiritual dimension into their designs, having first attempted to define the qualities desirable for this task. He quotes 'Abdu'l-Bahá's description of the character of the spiritually learned and the spiritual perfection which it implies as follows: 'The first attribute of perfection is learning and the cultural attainments of the mind . . .'.[60] The context of this counsel indicates that it was to promote an open-mindedness about all the different kinds of knowledge that may be obtained from different religions, cultures, nations and branches of knowledge; it therefore suggests a holistic approach. A builder's approach must entail not only a technical mastery of a profession, but a sensitivity to spiritual needs and an open and inquiring mind equipped to find creative solutions to the challenges involved in shaping the environment. There can be no room for thoughtless absorption of received wisdom and unquestioning acceptance of established ways of thinking, no matter how deserving of respect.

'Abdu'l-Bahá continues:

The second attribute of perfection is justice and impartiality. This means to have no regard for one's own personal benefits and selfish advantages . . .. It means to see one's self as only one of the servants of God . . . and except for aspiring to spiritual distinction, never attempting to be singled out from the others.[61]

This statement not only points past the over-emphasis on competition in building that Zrudlo observes; it indicates that 'learning' and 'cultural attainments' (relative as they are in certain respects from age to age) must at this time include a consciousness of the oneness of mankind.

Environmental Ethics and the Unity of Mankind: Education and Prosperity

The prerequisites for an architect's reaching his highest level of achievement are also the prerequisites for the people of the world, who are all called upon to create a continually advancing culture and to participate in the development, upkeep and protection of their physical environment:

> The third requirement of perfection is to arise with complete sincerity and purity of purpose to educate the masses: to exert the utmost effort to instruct them in the various branches of learning and useful sciences, to encourage the development of modern progress, to widen the scope of commerce, industry and the arts, to further such measures as will increase the people's wealth.[62]

Here we see recognition of the need for a degree of prosperity and stability as a basis for the healthy development of a community. 'Abdu'l-Bahá does not dismiss economic and social requirements. Indeed, he expresses a realistic awareness of these needs, with certain important provisos:

> Wealth is most commendable, provided the entire population is wealthy. If, however, a few have inordinate riches while the rest are impoverished, and no fruit or benefit accrues from that wealth, then it is only a liability to its possessor.[63]

Furthermore, 'Abdu'l-Bahá states that:

If a judicious and resourceful individual should initiate measures which would universally enrich the masses of the people,
there could be no undertaking greater than this.[64]

Note the emphasis on the word 'judicious': there is no room here
for the unscrupulous entrepreneur bent on exploiting the environment regardless of human or ecological costs. Furthermore, people
are not to be provided merely with material satisfactions—the
'bread and circuses' cynically offered to placate the Roman populace
huddled in the squalid insulae, the jerry-built tenement blocks
whose precarious conditions Juvenal so graphically describes.[65]
Rather, people should be given free access to the knowledge and
education that will help them to become active participants in the
creation of their environment. According to 'Abdu'l-Bahá, the built
environment is a part of the great whole:

We must now highly resolve to arise and lay hold of all those
instrumentalities that promote the peace and well-being and
happiness, the knowledge, culture and industry, the dignity,
value and station, of the entire human race.[66]

We see here an acknowledgement that such conditions are unlikely to arise without a true spiritual foundation. Earlier in the
same work, 'Abdu'l-Bahá gives us a clear statement of social mission,
stating plainly that material wealth must always come second to the
happiness and welfare of the people:

We should continually be establishing new bases for human
happiness and creating and promoting new instrumentalities
toward this end. How excellent, how honorable is man if he
arises to fulfill his responsibilities; how wretched and
contemptible, if he shuts his eyes to the welfare of society and
wastes his precious life in pursuing his own selfish interests
and personal advantages.[67]

The basis of a life devoted to the social good is an education that
teaches care for all beings:

O ye beloved of the Lord! The Kingdom of God is founded upon equity and justice, and also upon mercy, compassion and kindness to every living soul. Strive ye then with all your heart to treat compassionately all humankind except for those who have some selfish motive, or some disease of the soul. Kindness cannot be shown the tyrant, the deceiver, or the thief, because, far from awakening them to the error of their ways, it maketh them to continue in their perversity as before. No matter how much kindliness ye may expend upon the liar, he will but lie the more, for he believeth you to be deceived, while ye understand him but too well, and only remain silent out of your extreme compassion.

Briefly, it is not only their fellow human beings that the beloved of God must treat with mercy and compassion, rather must they show forth the utmost loving kindness to every living creature. For in all physical respects, and where the animal spirit is concerned, the selfsame feelings are shared by animal and man. Man has not grasped this truth, however, and he believeth that physical sensations are confined to human beings, wherefore he is unjust to the animals, and cruel.

And yet in truth, what difference is there when it cometh to physical sensations? The feelings are one and the same, whether ye inflict pain on man or on beast. There is no difference here whatever. And indeed ye do worse to harm an animal, for man hath a language, he can lodge a complaint, he can cry out and moan; if injured he can have recourse to the authorities and these will protect him from his aggressor. But the hapless beast is mute, able neither to express its hurt nor to take its case to the authorities. If a man inflict a thousand ills upon a beast, it can neither ward him off with speech nor hale him into court. Therefore is it essential that ye show forth the utmost consideration to the animal, and that ye be even kinder to him than to your fellow man.

Train your children from their earliest days to be infinitely tender and loving to animals. If an animal be sick, let the children try to heal it, if it be hungry, let them feed it, if it be thirsty, let them quench its thirst, if weary, let them see that it rests.

Most human beings are sinners, but the beasts are innocent. Surely those without sin should receive the most kindness and love—all except animals which are harmful, such as blood-thirsty wolves, such as poisonous snakes, and similar per-nicious creatures, the reason being that kindness to these is an injustice to human beings and to other animals as well . . .. Therefore, compassion to wild ravening beasts is cruelty to the peaceful ones—and so the harmful must be dealt with. But to blessed animals the utmost kindness must be shown, the more the better. Tenderness and loving kindness are basic principles of God's heavenly Kingdom. Ye should most carefully bear this matter in mind.[68]

This passage has been quoted at length because it expresses some of the most profound truths embodied in the Bahá'í Faith: com-passion, justice and humility are the basis of an education whereby we can bring our relationships with the physical environment together with the principle of the oneness of humanity. There is no room here for sentimentality; instead, we find a recognition that some creatures, because of their basic nature, constitute a threat to members of their own or other species, and require protective measures when their paths cross those of human beings or creatures under their protection and care.

It is possible, of course, to argue in specific instances that wolves, for example, cannot be regarded as 'bloodthirsty' in human terms, and that such a view of them represents an anthropomorphic impo-sition of human values.[69] In general, however, the principle remains valid: it is in keeping with the deepest tenets of Bahá'í belief to care for those who are weakest and most in need of succor and shelter, and specifically for those placed under one's own protection. Moreover, the Bahá'í believer is enjoined to pass on these precepts to his or her own children, allowing them from their earliest years to develop a relationship of trust and respect towards animals and to understand and meet their needs. The underlying message places the Bahá'í believer firmly within a multi-layered structure of inter-linking bonds of trust and tenderness, respect and reverence, in which there are endless lessons to be learned, often from those who seem to be the humblest in the family of creation.

Made up of 'the same constituent elements as man',[70] animals share our capacity for physical pain and are at the mercy of humans. Humans, in turn, are in the power of a Creator Whose mercy out-weighs our offences, a Creator Who admits us to a relationship not of tyranny and exploitation but of trust, confidence and love. The Creator offers us the gifts and resources to meet not only our physical but also our deepest spiritual needs. And so, as the child learns from the parent how to respond to the rest of creation and to meet the responsibilities of being human with justice and compassion, we, taking our patterns from God Himself, may come to develop the gifts which we have each received from Him. In doing so, we must claim the true dignity of humanity without pride or self-import-ance, and carry out His purpose for each of us and all of His creation.

From the passage cited above, it will be apparent that one quality above all others informs the activities of the builder, the architect and every other user of spiritual and material knowledge. In its deepest and simplest sense, this is nothing more nor less than love. Christian audiences will be familiar with Christ's summarizing of the Hebrew law and prophets in two all-embracing command-ments: to love God above all, and to love one's neighbor as oneself. In a similar way, the Bahá'í Faith involves an abiding love for one's fellow creatures which recalls the love of the Creator. Creativity, not destructiveness, is at the heart of this system of belief.

Environment Ethics and the Unity of Mankind: A World Federation

This need for a harmony with the created world that includes the prosperity of mankind and its advancement in knowledge becomes ever more urgent as pressures increase on the environment. Now that these pressures are overwhelming even the most advanced nations, there is a growing consciousness that the current *ad hoc* process for environmental legislation is lamentably insufficient. It cannot solve the most serious environmental problems within nations and it fails in addressing the more prevalent international environmental problems.

The case of Honduras, described earlier, presents a very real and tragic picture of the injustice in man's relationships with his fellow man and with the world of nature. Such current practices stand in stark contrast to the Bahá'í attitude of fairness and justice in the distribution of the fruits of the earth. These practices indicate, too, that the principle of the oneness of creation, spoken of in the many divers religious traditions of humanity, is today a dead letter. Current systems seem incapable of the educational, economic and legal measures required to adapt this principle to the new conditions of the world and establish it as a permanent element of the global society.

We return to 'Abdu'l-Bahá's words, cited earlier in brief:

O ye beloved of the Lord! In this sacred Dispensation, conflict and contention are in no wise permitted. Every aggressor deprives himself of God's grace. It is incumbent upon everyone to show the utmost love, rectitude of conduct, straight forwardness and sincere kindliness unto all the peoples and kindreds of the world, be they friends or strangers. So intense must be the spirit of love and loving kindness, that the stranger may find himself a friend, the enemy a true brother, no difference whatsoever existing between them. For universality is of God and all limitations earthly. Thus man must strive that his reality may manifest virtues and perfections, the light whereof may shine upon everyone. The light of the sun shineth upon all the world and the merciful showers of Divine Providence fall upon all peoples. The vivifying breeze reviveth every living creature and all beings endued with life obtain their share and portion at His heavenly board. In like manner, the affections and loving

kindness of the servants of the One True God must be bountifully and universally extended to all mankind. Regarding this,
restrictions and limitations are in no wise permitted.

Wherefore, O my loving friends! Consort with all the
peoples, kindreds and religions of the world with the utmost
truthfulness, uprightness, faithfulness, kindliness, good-will
and friendliness, that all the world of being may be filled with
the holy ecstasy of the grace of Bahá [Glory], that ignorance,
enmity, hate and rancor may vanish from the world and the
darkness of estrangement amidst the peoples and kindreds of
the world may give way to the Light of Unity. Should other
peoples and nations be unfaithful to you show your fidelity
unto them, should they be unjust toward you show justice
towards them, should they keep aloof from you attract them to
yourselves, should they show their enmity be friendly towards
them, should they poison your lives, sweeten their souls,
should they inflict a wound upon you, be a salve to their sores.
Such are the attributes of the sincere! Such are the attributes of
the truthful.[71]

According to 'Abdu'l-Bahá, unity depends greatly upon individual character and the acquisition of virtues. Man's responsibility
to his fellow man is a spiritual one, is divine in nature, and is
essential to the prosperity of humanity as a whole. This call for
harmony between peoples and nations is the prelude to a call for
the formation of social instruments that can reinforce this harmony
and ensure that its consequences are justice and the best interests of
mankind as a whole, both now and in the future.

When the individuals of the world fulfil their responsibility to
love, care for and serve their fellow creatures and fellow men, they
will have laid down the bedrock for a spiritual world order. 'Abdu'l-
Bahá outlines the broad features of an 'all-embracing Pact',[72] to be
agreed upon as a first step towards the unity of nations, 'causing all
the peoples of the world to regard themselves as citizens of one
common fatherland'.[73] On this basis, 'the world's federated repre-

sentatives' can build up the laws and systems prerequisite for the development of a practicable and refined environmental ethic that can be applied internationally.[74]

The inefficient, fragmented and unsystematic way of addressing the environment that is presently in use should therefore be replaced, in a Bahá'í view, by a world federal system operating through a world parliament. A wider loyalty to an earth that 'is but one country, and mankind its citizens',[75] accompanied by 'a single code of international law',[76] would alleviate the problem of waiting for states to sign, ratify or update legislation in order to comply with their particular laws.[77] This sense of belonging to one earth would generate new thinking about the mutuality of the physical and spiritual activities of which the planet is the joyous and holy site; it would revitalize the ancient principle of the oneness of creation, and would make the precept of brotherly love a reality among the whole human family. Thus the new spiritual order would nurture, in institutions, communities and individuals, the prerequisites of an effective environmental ethic.

JEREMY NAYDLER
The Three Temptations

There is much talk today about the relative influence of genetic and environmental factors on the formation of human nature. But human beings are also spiritual beings with an inner life of soul. Our spiritual impulses are every bit as important as the physical drives and influences that affect us both from within and without.

One important stream of thought within Western metaphysics is the Hermetic tradition, which has a two thousand year history. In the Hermetic tradition there is a treatise known as the *Asclepius*, where we find a beautifully succinct statement concerning the purpose of human life:

> God made human beings out of the substance of spirit (*animus*) and the substance of body (*corpus*)—of that which is eternal and that which is mortal—blending and mingling together portions of each substance in the right measures, in order that the creature so fashioned might be able to fulfil the demands of both sources of its being: that is, to venerate and worship the things of heaven, and at the same time to tend and care for the things of earth.[78]

According to the *Asclepius*, because we are not just physical creatures but have also a spiritual dimension to our nature, we cannot fulfil ourselves as human beings if we neglect this spiritual dimension. But neither would we be fully human if we were pure spirits. We exist between the poles of spirit and matter, and that is and always has been the wonder and struggle of being human. For our nature is such that there are two tendencies in us that work in opposite directions. On the one hand, there is a tendency to neglect the fact that we are spiritual beings as well as physical beings, in which case we may fail 'to venerate and worship the things of heaven', but spend our time fulfilling our earthly desires. On the other hand, it is also possible to neglect the fact that we really are creatures of the earth, and to think that we don't belong here. Then we fail 'to tend and care for the things of the earth'.

The recognition of these two pulls in our nature runs through the Hermetic tradition. It can also be found in Aristotle's *Nicomachean Ethics*, where the central teaching is that the life of virtue is lived between two extremes. Aristotle goes through a long list of virtues to show how each of them is positioned as the mean between the opposed tendencies to disregard either the spiritual or the physical aspect of the human makeup. The virtue of courage, for instance, is between rashly exposing yourself to danger (disregarding the physical) and a cowardly saving of your own life at all costs (disregarding the spiritual).[79] Aristotle was concerned with how to be human, how fully to realise our nature as spiritual beings living on earth.

Aristotle's doctrine of the mean is echoed in the Buddha's Middle Way. It is also something we meet again in the temptations experienced by Christ before he started his ministry, as these are recorded in the Gospel of Matthew.[80] If we regard Christ as the divine human archetype, then we could say that the temptations that he undergoes in the wilderness are archetypal, in the sense that they apply to all of us as individuals. They are also peculiarly relevant to our present cultural situation, for they involve a battle for the true human image, and this is the great battle that is being fought at the beginning of the twenty-first century. So I want first briefly to remind you of the three temptations, and then I will try to show why I think they have a special relevance to our cultural situation today.

The first of Christ's temptations was the temptation to turn stones into bread. This arose when he was very hungry, and struggling against the physical demands of his body. The temptation was essentially to turn away from his spiritual nature, and to focus all his powers on transforming matter for his own benefit. His answer to the tempter was therefore: 'Man does not live by bread alone . . .'.[81]

Christ is then taken to the pinnacle of the Temple in Jerusalem, and told that if he casts himself down the angels will bear him up. This can be understood as a temptation to turn away from the physical aspect of his being, defying the laws of gravity. It is a temptation *not to engage* with the reality of the material world.

In the third temptation, he is taken to a high mountain and shown 'all the kingdoms of the world' in their glory, and told he can have them all if he worships Satan. This temptation is of a different order, because it is to do with his whole life's purpose. What had he come into the world to do? What was his real aim in life—simply to have power over the world, or to fulfil a spiritual purpose? Jesus's answer is that we should 'worship and serve God': that is the ultimate purpose.

I want now to look at current cultural trends in the light of each of these three temptations of Christ, because I think that we are today facing these three temptations collectively. In case this begin to sound like a sermon, I want to make it clear that it is not my intention to speak about Christian doctrine, but to focus rather on a profound truth about the human condition, and the fact that we are today being tested as never before concerning what it really means to be human. In this respect we may think of Christ both as the representative of humanity, and as our highest potential, embodying what it means to be truly human.

The First Temptation

The first temptation, to turn stones into bread, is a temptation to turn away from our own spiritual nature and spiritual needs, and to focus all our powers on manipulating matter for our material benefit. This has been one of the driving forces behind materialistic science since the seventeenth century, and it first revealed itself in the desire to explain everything in terms of material causes. Recourse to God would then become unnecessary. Thus we have the statement of the eighteenth-century mathematician and physicist Laplace, reportedly made in conversation with Napoleon, that God is 'an unnecessary hypothesis'.[82]

The spiritual impulse doesn't go away, though. It simply gets displaced. From the beginning there was a kind of religious zeal driving the inaugurators of the scientific revolution—people like Bacon and Hobbes. And one finds it in modern scientists too—Francis Crick, Jacques Monod, Richard Dawkins, and many others who want to prove there is nothing more to life than physical processes.

Back in the seventeenth century, it was still felt necessary to give a place in the brain to the soul, which Descartes thought resided in the pineal gland. Today the soul, like God in the seventeenth century, has also become an 'unnecessary hypothesis'. The soul is replaced by the more intellectual concept of 'mind', and the mind is regarded as a 'neural computer'.[83] Today people talk about 'left brain' and 'right brain' thinking, as if logical thinking or intuitive, creative thinking are actually brain functions, and the person who is doing the thinking or being creative doesn't exist. It is quite alarming *to what extent people are unconsciously adopting a way of speaking about the soul in terms that actually exclude it.*

And so the true image of the human being is lost. Meanwhile the physical body is regarded as a machine. I have a biology textbook for children written in the early 1990s called *The Human Machine,* whose introduction contains the following statement: 'The human body is a fascinating and remarkable machine. Its design is far more complex than the most advanced computer . . .'.[84]

This is a standard textbook for children! Turn the pages, and there is a picture showing the different joints in the human skeleton in the clean lines that one would find in any manual of basic engineering: pivot joint, hinge joint, ball and socket joint, etc.

If the body is a machine, then it can be treated as one would treat a machine. When parts wear out, then you replace them with new parts. From Bacon onwards, the avowed main purpose of scientific research was to discover things useful for human beings in order 'to subdue and overcome the necessities and miseries of humanity'.[85] Again and again this is put forward as the main justification for scientific research today. But this alliance between purely materialistic thinking about causes, and a utilitarian view of the purpose of knowledge, presents a grave challenge to the true image of the human being. Let me give you an example.

It is now proposed that once a person dies their organs should be made available to anyone else that needs them.[86] At the moment you have to have a donor card if you are prepared to let your organs be used by someone else. The new proposal—if it became law—would legally establish the principle that the human body is just a conglomeration of parts that are unensouled, and can therefore be removed, swapped or replaced by someone else's equally unensouled body parts.

As it is, there is already a massive multi-million dollar trade in 'body parts'. In 1997, in the USA alone, the market for human organs was worth $6 billion.[87] No doubt it is worth considerably more today. The recent scandal in which a huge collection of organs was built up at Alder Hey children's hospital in Liverpool is a good example of how entrenched this mentality is: the medical profession were taken by surprise that some people still regard body parts as in some sense infused with the soul of the person to whom the body parts belonged. According to a report published at the end of January 2001, there are over 100,000 body parts kept in storage in British hospitals and medical schools.[88]

There was an interesting case during the summer of 1999: a teenage girl was forced against her will to have a heart transplant. She said that she did not want someone else's heart and would prefer to die.[89] In any other period in history, the very idea of heart transplants would have been considered diabolical. It rests on the premise that a person's identity is located in their brain—which is a very modern idea. This particular case raises a troubling question. In scientific terms the girl's life was saved, but her heart was cut out of her body and someone else's heart is now in her body: does she feel that she is the same person? There are some disturbing accounts of heart transplant patients who have undergone personality changes, taking on the personalities, likes and dislikes of the heart donors, so this is a question of real concern.[90]

The direct corollary of transplant surgery is the fantasy of replacing the body altogether by a machine. It is a suggestion that has been seriously proposed by respected scientists throughout the twentieth century—from J. D. Bernal (physicist and historian of science) in the 1920s to Hans Moravec (professor of engineering at MIT) in the 1990s. The fantasy is that an operation could be performed to extract the brain and attach it to the latest model of computer, or computerized robot. The assumption is that the brain alone holds the person's identity and the rest of the body is wholly expendable. We then incarnate in a 'post-biological' body of cold metal and electronic circuitry.

The great advocate of this today is Kevin Warwick, Professor of Cybernetics at Reading University. As he says: 'In this way we would be able to give ourselves a type of immortality, with not only our body being replaced by a new gleaming robot body, but also our brain being replaced by a new, much more powerful, faster, more accurate machine brain.'[91]

One of the supposed advantages of the new machine-human hybrid (the cyborg) is that it would abolish the need for sleep, and enable us to be twice as industrious. So it isn't just the human organism that is being attacked here, it is the soul and those vital domains of the soul's life—sleep, dream and spontaneous feeling. So far we can treat it simply as a fantasy, but the prospect of performing successful brain transplants may not be so far down the road.

Another expression of this same assault on the integrity of the human being is the more realistic possibility of replacing human organs with organ transplants from genetically modified animal donors, such as pigs. This is very close to being practised: probably just three years away. The first cloned pigs were born last year, and it is no secret that the main reason for doing it is to produce 'organ farms' for pig-human transplants.[92] Not only are the pigs to be genetically modified, but the human recipients will be given transfusions containing modified pig cells. At the end of 1999, new guidelines were announced that would require anyone who received one of these organs to sign a declaration that they would not have children.[93] One shudders to think what might emerge into the world if they did.

But the future of transgenic organ transplants does not just constitute an assault on the human being. It is also the pig who is assaulted by a purely secular and utilitarian consciousness that does not acknowledge the integrity of the animal itself. It fails to see that nature is sacred, that the divine extends right into the physical world. The practice of genetic engineering is absolutely relevant to the wider question of our whole relationship to the spiritual dimension of existence, our relationship to the divine world, which is the source and ground of the natural world. According to Goethe, 'God is operative in nature and nature in God, from eternity to eternity'.[94] Our relationship to nature, then, is at the same time our relationship to God.

In January of this year, scientists successfully created the world's first genetically modified monkey, by transferring jellyfish genes that make its fingernails and toenails glow.[95] Of course this was done for medical research purposes, so we are all supposed to think it perfectly justified. There now exists the possibility of redesigning animals, plants and also humans; and indeed this is already being done on a massive scale. A Government report, published last July, revealed that in 1998 just under half a million animal experiments involving genetic modification were performed in the UK.[96] There is usually some compelling medical reason given, but what lies behind such experiments is a purely materialistic view of the evolution of species, one which is blind to the spiritual archetype of the animal. A violation of the animal's essential integrity is then not seen as a violation. It becomes justifiable.

But is it? There are literally tens of thousands of transgenic mice used in our laboratories, nearly all for medical research.[97] The 'oncomouse', for example, is specially designed so that it is prone to develop cancer. But can we truly say that our consciences are clear in creating a creature designed to suffer on our behalf? The increasing number of cases of the commercial genetic modification of animals shows even more clearly the ethical dubiousness of our 'playing God' in redesigning creatures for our own ends. We are now, it seems, at the threshold of repopulating our farms with transgenic stock: a new 'low fat' pig with lower than normal fat (this was achieved by introducing human and mouse genes into the pig's DNA), a new breed of featherless chicken, a new salmon that grows six times as fast as normal, and cows that produce human breast milk.[98]

While the modification of plants perhaps seems a less emotive issue, who does not feel just a twinge of sadness at the prospect of a new 'improved' oak tree that grows faster and straighter than the old type that has graced our woods for millennia?[99] Is it just sentimentality that tells us that the essential character of the oak—what makes it an oak—is that it grows so slowly and never so straight as a pine? Are we not undermining its essential nature by forcing it to grow differently? Do we have the right radically to alter natural organisms just because it is of apparent benefit to us to do so? For the most part—and this includes medical research—genetic engineering is driven by big business, with the primary motivation of making big money. For example, the cloning of transgenic pigs will be hugely profitable.

This is just the beginning of an all-out assault on the spiritual integrity of animals and plants. It is based on the treatment of nature as a human resource, with no intrinsic value and no intrinsic rights. I would suggest that it could only take place within a world view that is not just Godless but in a real sense diabolical, and in which human beings have cultivated a kind of diabolical cleverness.

It is likely that genetic research will eventually produce new species that fall half-way between human and animal, as one scientist has said: 'There is no logical or practical reason why we cannot give human genes to chimpanzees with the aim of giving them a spoken language . . .. Perhaps we could use them as intelligent subhuman clones for difficult and dangerous tasks, instead of incredibly expensive and limited robots.'[100]

What is it that is driving science to think in this way? As the title of one of Goya's more nightmarish pictures puts it, 'The dream of reason produces monsters'. While it may seem to be a dream of becoming like God, one actually ends up becoming more like the devil.

The story of Frankenstein and the monster that he created is surely the presiding myth of twentieth- and twenty-first-century science. It is ironic that the name of the monster's creator has now become virtually synonymous with the monster itself. When the name 'Frankenstein' is mentioned, everyone thinks of the monster, as if in this confusion of identities a deeper truth is revealed: that the more monstrous new forms we create, the more monstrous will our own nature have become.

Behind this drive to desacralize and to desecrate both nature and the human being we see the temptation 'to turn stones into bread'. We need to know the nature of the being who tempts us in this way. In Matthew's account he is called the *diabolos* or 'devil': the root meaning of *diabolos* is 'the opposer'.[101] Let us say he represents tendencies at work within human nature that oppose our true fulfilment. But the *diabolos* has at least two aspects, corresponding to two quite different tendencies. In this first aspect, tempting Christ to turn stones into bread, he is usually portrayed by artists as ugly, deformed or serpentine. What he seeks to do is to drive out any residual awareness that God is present in nature, or that the human being is in any way capable of reflecting the divine. Rather, he would have us drift further and further from awareness of spiritual realities into a subearthly, subhuman realm where everything is ugly and deformed. His aim is to tempt us to become so monstrously clever that we will lose our humanity.

The Second Temptation

The second temptation of Christ is of a quite different nature from the first. It is to ignore the reality of the material world and to turn away from any real engagement with it. In the United States, there exists within certain fundamentalist groups the recurrent idea that the Last Judgement is going to happen at any moment, and that the believers will be transported instantly to heaven, leaving everyone else behind to a miserable fate as the world descends into total chaos. The word 'rapture' is used to describe this dream, which is a symptom of the widespread fantasy of *abandoning the earth*.

In the 1990s there were several mass suicides by members of small cults, for example Heaven's Gate (1997, in California), and Solar Temple (1994, in Europe and Canada). These groups share the belief that our true human destiny is not really on earth at all, and that by committing suicide they release their spirit to travel to heaven (in the case of Heaven's Gate, by flying saucer). This is not just a phenomenon restricted to the global North. There are a lot of fundamentalist Christian cults in central Africa. Three hundred members of one such cult in Uganda, called the Movement for the Restoration of the Ten Commandments, committed mass suicide last year believing that the Virgin Mary would carry them to heaven.[102]

More seriously, and certainly more expensively, there is the powerful technological fantasy of abandoning the earth in a spaceship or dressed in a spacesuit.[103] This is essentially a dream of disincarnation symbolized by the new technological shaman, the astronaut. The astronaut abandons the body as well as the earth. For he does not breathe real air, does not eat fresh food, does not experience the sensation of warm sun on his skin, does not experience the joy of walking, of hearing or touching anything directly.

It is extraordinary to contemplate the fact that plans are already underway to build a space hotel that should be up and running by the year 2020. It is called Hotel Galactica and is being designed by Kawasaki: 6000 Germans have already made reservations (costing £300 each).[104] Whether or not this happens, the project itself is an expression of the desire to desert the real world for the virtual world.

It would seem that this desertion has already reached epidemic proportions in contemporary society. It is a fact that after sleeping and working, the third largest use of time in the USA and Europe is watching television. In 1999, people in the UK spent an average of 26 hours a week watching TV. That is more than 3½ hours each night *on average!*[105] But what happens when we watch TV? The soul is drawn out of itself, to inhabit an inner world that is not our own. Instead of putting us in touch with what is going on in our own soul-life, people are lured into a manufactured, artificial inner world.

But TV is nothing compared to the possibilities opened up by interactive virtual technology, where one can have the experience of literally entering virtual space. By the 1980s, it was clear that the very term 'virtual world' was to signify a new reality. This reality came into its own with the development of the 'stereoscopic head-mounted display' (first developed, interestingly enough, by NASA), that gives one the impression of being totally surrounded by virtual 3D space. Add to this 'data gloves', and you have the sensation of touching things that exist only virtually. It is now possible to wear full body suits (aptly termed 'technological envelopes'); and people in different parts of the world can explore the same virtual reality through a telecommunication link-up. Escape from the here and now of one's physical existence is thus complete.

Of course virtual experiences, like drugs, are particularly attractive to adolescents because one is transported into a fantasy world. What you experience virtually is so much more amazing than what you experience in ordinary consciousness. I believe that part of the allure of virtual experience, and this applies to the taking of drugs too, is that it serves as a substitute for genuinely mystical experience. Whether that is the case or not, however, it is certainly a substitute for attending to, and caring for, the actual world in which one lives.

Meanwhile the real world becomes increasingly polluted and increasingly dangerous to live in. Even sunshine has become a source of peril. But while we have only ourselves to blame for the degradation of the environment, it seems that most of us are highly susceptible to the temptation simply to beat a retreat to the safety of virtuality. This retreat from the real can take a host of different forms, all now regarded as completely normal: listening to the walkman or talking to a friend on the mobile while walking down the High Street, objectifying the world through the camera lens, travelling at high speed in a motorcar, watching a video, surfing the Net . . . .. The list of such activities in which we are 'elsewhere' could be extended almost indefinitely, they are so ubiquitous. In a sense, we have all become astronauts now, only half-present on the planet.

Behind this impulse, to turn away from the earth and embrace the virtual world, lies the temptation to cast oneself from the pinnacle of the Temple in a gravity-defying gesture. The aim of the *diabolos*, as it is described in Matthew's Gospel, is to undermine Christ's sense of really belonging to the earth. Christ is tempted to indulge in a false spiritual experience of 'flying', believing that as a spiritual being he has no need to defer to physical reality. Whereas the aspect of the *diabolos* that would have us turn stones into bread is usually portrayed as ugly and deformed, this aspect is usually depicted by artists as exceedingly beautiful. For here we are dealing with the fallen angel Lucifer, who tempts us away from all engagement with matter and material reality.

There is a particularly revealing story concerning Lucifer, and how it was that he came to fall. It is related that when God first created Adam, He invited all the angels to come and worship His new creation, for into Adam God had instilled a spark of His own divine nature, and He was very pleased that He had been able to create a creature of the earth, yet at the same time divine like Himself. And all the angels came and bowed down before Adam. But when Lucifer came, he proudly refused to bow before Adam, saying 'What? Am I to bow before this thing of clay! Never!'[106]

In Lucifer's nature, then, there is a disdain for the earth, and a hatred of the fact that the human spirit is incarnate in a physical body. For him the earth cannot harbour anything of value. He promises wonderful and mind-blowing experiences—but only if we join him in spurning the earth.

The Third Temptation

We have now considered the first two temptations—one to turn away from our spiritual nature and practice a 'black alchemy' on the material world; the other to flee engagement with the material world so as to lose ourselves in counterfeit spiritual experiences. The third temptation is different from both of these. It is neither to cut off from the spiritual or the material pole, but rather to cut off from the centre, from the connection with the deepest wish that one can have as a human being.

For Christ, the third temptation was the temptation of power, a self-aggrandizement that serves only the ego and not the true self. But there are other things that it could have been—fame, for example, possessions, wealth or some other false security. People are so susceptible to misidentifying what they really want, thinking that any number of things—if only they possessed them—would guarantee their 'happiness'. In a way this is the most testing of all the temptations, because it requires that we be fully awake. And unless we are totally sure what our ultimate goal in life is, and what we value and cherish above all other things, we remain vulnerable to this temptation.

As an example of this vulnerability, let me cite the disturbing fact that in 1999 the sale of pornographic 'adult films' worldwide topped $10 billion.[107] This is an indication of just how susceptible we are today to having our energies and our focus diverted from our deeper human values. The pervasiveness and 'acceptability' of pornography (hotel chains in the USA seem now to offer adult films to their customers as a matter of course) is a general symptom of a massive failure to connect to our real values and purpose as human beings. For the hidden agenda of pornography, by tearing the human sexual act from its emotional and relational context, is to persuade people that there is no such thing as the human soul or spirit, only the animal body. In this respect there is an implicit alliance between pornography and reductionist science. Both are intent on denying the reality of the human spirit.

The global pornography industry is just one very obvious example of how our essential dignity as human beings can be undermined. But the third temptation comes into operation in more subtle ways on a daily, even hourly basis, to the extent that we fail to reconnect with our own centre. From a spiritual perspective, modern civilization could be described as being profoundly 'off-centre'. The modern world is replete with any number of distractions that would take us away from experiencing the 'one thing' that alone can give our lives meaning and purpose. We busy ourselves with phone calls and shopping, and then there are the newspapers to read, the emails to check, the TV programmes to watch. For most of us, even when we make resolves to use our spare time to meditate or to reflect on our dreams, it is extremely difficult to resist the thousand distractions that we ourselves line up against allowing ourselves to dwell in our own inner space.

In Matthew's account of the third temptation, Christ is led to a high mountain, but it was not just so that he could survey the view. It was, I suspect, so that he would experience being right at the edge of an existential precipice. For this temptation has to do with each one of us coming up to the edge of an inner abyss or void, and experiencing the panic that would make us grasp at straws, anything to make us feel secure. The challenge, though, is to stay there until we know what it is that we really have to do with our lives.

Modern secular society, with its technological and industrial power and ingenuity, is largely driven by the terror of this inner abyss. We do not simply 'live in' a secular society, it is something we are all constantly creating and sustaining in order to avoid experiencing inwardness. And yet we cannot be truly effective in our lives unless we have come to know and live with that experience. It may be uncomfortable, it may even be terrifying, but to allow ourselves the experience of dwelling in our own space rather than surrendering to the next distraction is to make a step toward the resacralization of our lives, and to begin a truly radical transformation of the secular society. As the British philosopher A. N. Whitehead has pointed out, the first requirement of the religious life is to experience solitariness. 'Religion', he wrote, 'is what the individual does with his solitariness. If you are never solitary, you are never religious.'[108] In modern urban industrial society, solitariness is viewed like the plague, to be fled from, rather than to be embraced as an opportunity for spiritual deepening.

In the recently published report of the UN International Panel on Climate Change, the most authoritative that has so far been produced, it is predicted that the effects of rising global temperatures in the coming century are likely to be catastrophic for the world.[109] Today we all face the very real prospect that irreversible changes have already occurred which are likely to lead to disasters on a scale unprecedented in recorded human history. We now have good reason to assume that through the effects of human-induced global warming, modern civilization will come to an end within the next few generations. Let us suppose this to be the case. Let us suppose the death sentence has already been passed on our civilization: it is now just a question of how long it takes to be executed. What, then, should be our response as individuals, given that we are more or less powerless to make any appreciable impact on the process that is already well under way? How should we live our lives?

There is a story that the Buddha told, of a famous and stunningly beautiful dancer who had just come to a small town. People were swarming the streets, eager to catch a glimpse of her. At that same moment, a condemned criminal was obliged to cross the town square carrying a bowl of oil that was filled to the very brim. He had to concentrate with all his might on keeping the bowl steady, because if one drop of oil were to spill from the bowl to the ground, the soldier directly behind him had orders to take out his sword and cut off the man's head.

Having reached this point in the story, the Buddha asked: 'Now do you think the prisoner was able to keep all his attention so focused on the bowl that his mind did not stray to steal just a glimpse of the beautiful dancer in town, or to look up at the throngs of people who were all around him and who might at any moment bump into him?'[110]

The Buddha is succinctly describing the response that we need to make to the third temptation. Like the criminal, we do stand a chance of coming through, but it will require tremendous dedication, tremendous effort to keep faith with our core values. What is being challenged today is above all else our sense of what it means to be human. And that requires that we each of us discern what it is that ultimately leads to the fulfilment of our real purpose.

SEYYED HOSSEIN NASR
The Spiritual and Religious Dimensions of the Environmental Crisis[111]

❧

Considering the depth and breadth of the environmental crisis and the heedlessness of those who continue to pursue the very means and abet the very forces which have brought the crisis about, it might appear to be futile to speak about it again and again. But this crisis has a spiritual and religious dimension and is the result of the forgetting of certain perennial truths which need to be asserted and reasserted amidst the chaos in which we live. To express the truth is in fact the most important of all acts and one should take every opportunity to do so even if it seems to have no effect, at least none that is perceptible. Even if we are not able to perceive the effect, however, surely the expression of the truth bears its fruit; otherwise it would not have been considered as such a virtuous act in various traditions.

In any case there is nothing more timely to discuss than the question of the environmental crisis and the truths and falsehoods associated with this whole matter. It is not accidental that the word crisis is used in this context, for a veritable crisis it surely is, following upon the wake of that spiritual and intellectual crisis which is inseparable from the very world-view of the modern world. The earlier crisis which René Guénon discussed three quarters of a century ago in a number of works, including *Crisis of the Modern World*,[112] was known to the few and ignored by the many. The environmental crisis, however, is too manifest to be ignored even by the multitude. It is a crisis of the utmost gravity and urgency and anyone who neglects it is simply fooling himself or is daydreaming. It is, however, in our nature to try to evade confrontation with what requires of us the deepest transformation within ourselves.

The subject for my talk tonight has been chosen by my hosts. We have spent the whole day today here in London talking with some of the leading British theologians and scientists about this same question of the environmental crisis and some of them are in this audience tonight. Probably they are tired of it being discussed again, but because of the urgent nature of the subject, I am always glad though saddened to speak about it and to repeat certain basic theses. The situation today reminds one sometimes of the last few hours of the *Titanic* when people were playing music while their ship was sinking. We are obliged to turn again and again to these matters since, unfortunately, the negative forces which are bringing chaos upon our world continue unabated. In fact, it is necessary to repeat the theme of the environmental crisis even if it seems that no one is listening because, as already mentioned, the very assertion of the truth on any level is itself the most positive of all acts. Even if outwardly it seems to have no effect, in reality it does. Even if one person were to change his or her view that might mean a great deal.

It may be our nature to try to evade an imminent danger unless we really face it, but we do not want to face it for the very reason that it is a danger. The grave picture that is painted by serious scholars and honest scientists who are interested in the future of humanity can often be counteracted by having a film company send a camera into the woods to photograph a few birds flying around with the pretense of showing how 'normal' the environmental situation of the earth is even in urban areas. The truth is, however, otherwise. There is a major crisis at hand and it must be taken completely seriously. Moreover, one must also realize that the environmental crisis cannot be solved by good engineering (or better engineering), cannot be solved by economic planning, cannot even be solved by cosmetic changes in our conception of development and change. It requires a very radical transformation in our consciousness, and this means not discovering a completely new state of consciousness, but returning to the state of consciousness which traditional humanity always had. It means to rediscover the traditional way of looking at the world of nature as sacred presence.

For the title of my lecture, as you can see, I have chosen both the words 'spiritual' and 'religious'. That was done on purpose, because the present usage of the word religion in many quarters often leaves out precisely the spiritual element. Those people who are looking for the inner dimension of religious experience and of religious truth, are seeking another word to supplement the word religion. It is tragic that this is so but it is nevertheless a fact. The word spirituality in its current sense is a modern term, and not the Latin term from which it derives. As far as my own research has shown, the term spirituality as it is used today began to be employed by French Catholic theologians in the mid-nineteenth century and then crept into English. We do not find the use of this term as we now understand it earlier than the last century. Today it denotes for many people precisely those elements of religion which have been forgotten in the West and which therefore have come to be identified wrongly with spirituality as distinct from religion. From my point of view which is always of course a traditional one, there is no spirituality without religion. There is no way of reaching the spirit without choosing a path which God has chosen for us, and that means religion (*religio*). Therefore, the reason I am using both words is not for the sake of expediency, but to emphasize that I mean to include a reality which encompasses both spirituality and religion, in the current understanding of these terms, although traditionally the term religion would suffice since in its full sense it includes all that is understood by spirituality today.

It is important that we remember that all of us on the globe share in destroying our natural environment, although the reasons are different in different parts of the globe. In the modern world the environment is destroyed by following the dominating philosophy, while in what remains of the traditional world it is done in spite of the prevailing world-view and most often as a result of external coercion as well as temptation, whether it be direct or indirect. I have repeated this truth in many places and have caused some people to become angry, but the fact is that the only action in which nearly everybody participates at the present moment of human history, from communist to socialist, to capitalist, from Hindu and Muslims to atheist, from Christian to Shinto, is in living and acting in such a way as to cause the destruction of the natural environment. This fact must seep fully into our consciousness while at the same time we remember the differences in motive and perspective among religious and secularized sectors of humanity. Obviously, for those for whom religion is still a reality, it is much easier to appeal to religion and the religious view of nature to discover the means through which a solution would be found for the crisis from which we all suffer.

We often forget that the vast majority of the people in the world still live by religion. And yet most Western intellectuals think about environmental issues as if everyone were an agnostic following a secular philosophy cultivated at Oxford, Cambridge or Harvard and so they seek to develop a rationalist environmental ethics based on agnosticism, as if this would have any effect whatsoever upon the environmental crisis. It is important to consider in a real way the world in which we live. If we do so then we must realize why in fact religion is so significant both in the understanding and in the solution of the environmental crisis. Let us not forget, I repeat, that the vast majority of the people in the world live according to religion. The statistic that is often given, saying that only half of humanity does so, is totally false because it is claimed that in addition to the West one billion two hundred million Chinese are atheists or non-religious. This is not at all the case. Confucianism is not a philosophy, but a religion based upon ritual—I shall come back to that in a few moments. There are at most a few hundred million agnostics and atheists spread mostly in the Western world, with extensions into a few big cities in Asia and Africa. But this group forms a small minority of the people of the world. Those who live on the other continents, as well as many people in Europe and America, still live essentially in a religious world. Although in the West the religious view of nature has been lost, even here it is still religion to which the most ordinary people listen, while the number is much greater in other parts of the globe. That is why any secularist ideology that tries to replace religion always tries also to play the role of religion itself. This has happened with the ideology of modern science in the West which for many people is now accepted as a 'religion'. That is why the people who try to sell you many kinds of goods on television do so as 'scientists'—as agents of 'authority'—and always wear a white robe, not a black robe of traditional priests. They are trying to look like members of the new 'priesthood'. They function as the 'priesthood' of a pseudo-religion. Their whole enterprise is made to appear not as simply ordinary science but as something that replaces religion. For people who accept this thesis it would be feasible to accept a rationalistic ethics related to science but the vast majority of the people in the world still heed authentic religion. Consequently, for them, no ethics would have efficacy unless it was religious ethics.

In the West, for four hundred years, philosophers influenced by scientism have been trying to develop secular ethics and, sure enough, there are many atheists who are very ethical in their life, but by what norm are they to be considered as ethical? By no other than the very norms which religion instilled in the minds of people in the West. If somebody murders his neighbour we think it is unethical. But why is it unethical? What is wrong with that? The television programmes you watch on nature in Africa show that animals are eating each other all the time. If we are just animals, then what is wrong if we kill one another? The fact that everybody says 'no' to such an act is precisely because there are certain religious values instilled even into the secular atmosphere of the modern West which speaks of so-called secular ethics. The values of this ethics really have their roots in religion. In any case no secular ethics could speak with authority except to those who would accept the philosophical premises of such ethics.

The fact remains that the vast majority of the people in the world do not accept any ethics which does not have a religious found-ation. This means in practical terms that if a religious figure, let us say, a mulla or a brahmin in India or Pakistan, goes to a village and tells the villagers that from the point of view of the Sharī'ah (Islamic

Law) or the Law of Manu (Hindu law) they are forbidden to cut this tree, many people would accept. But if some graduate from the University of Delhi or Karachi who is a government official comes and says, for rational reasons, philosophical reasons, that it is better not to cut this tree, few would heed his advice. So from a practical point of view the only ethics which can be acceptable to the vast majority, at the present moment in the history of the world, is still a religious ethics. The very strong prejudice against religious ethics in certain circles in the West, which have now become concerned with the environmental crisis, is itself one of the greatest impediments to the solution of the environmental crisis itself. This fact cannot be doubted in any way.

There is a second reason why religion is so important in the solution of the environmental crisis. There are many elements involved here but as this is only a one hour lecture I have to summarize. We all know and even if we are not personally concerned with the metaphysical, spiritual and cosmological roots of the environmental crisis, we are none the less aware of the fact that outwardly (I do not say inwardly) this crisis is driven by the modern economic system appealing to human passions, especially the passion of greed intensified by the creation of false needs, which are not really needs but wants. This is in opposition to the view which religions have espoused over the millennia, that is, the practice of the virtue of contentment, of being content with what one has. The modern outlook is based on fanning the fire of greed and covetousness, on trying to do everything possible to attach the soul more and more to the world and on making a vice out of what for religion has always been a virtue, that is, to keep a certain distance and detachment from the world; in other words a certain amount of asceticism. There is a famous German proverb 'there is no culture without asceticism'; and this is true of every civilization.

We are living in the first period in human history in the West in which, except for a few small islands here and there of Orthodox or Catholic or Anglican monasticism, and a few people who try to practice austerity, asceticism is considered to be a vice, not a virtue. It is not taught in our schools as a virtue; it is taught as a vice, preventing us from realizing ourselves, as if our 'selves' were simply the extension of our physicality. This idea of self-realization is of course central to the Oriental and certain Occidental traditions. But it has become debased in the worst way possible today and transformed into the basis for modern consumerism, which can be seen in its most virulent form in America—now fast conquering Europe, and doing a good job of reaching India, China, Indonesia, etc. (within the next decade we will have several billion new consumers in such countries thirsting for artificial things which they have lived without for the last few thousand years). And what this will do to the earth God alone knows. It is beyond belief and conjecture what will happen if present trends continue. So what is it that can rein in the passions, either gradually or suddenly? Nothing but religion for the vast majority of people who, believing in God and the afterlife, still fear the consequences of their evil actions in their lives in this world. If it were to be told to them that pollution and destruction of the environment is a sin in the theological sense of the term they would think twice before indulging in it. For the ordinary believer the wrath of God and fear of punishment in the afterlife are the most powerful force against the negative tendencies of the passionate soul. For nearly all people on the earth who continue to pollute the air and the water, and whose life style entails the destruction of the natural environment, what is it that is going to act as a brake against the ever-growing power of the passions except religion? The religions have had thousands of years to deal with the slaying of the passionate ego, this inner dragon, to use the symbol mentioned in so many traditions. St Michael's slaying of the dragon with his lance has many meanings, one of which of course is that the lance of the Spirit alone is able to kill that dragon; or what in Sufism is called *nafs*, that is, the passionate soul, the lower soul within us. We rarely think of that issue today. But where is St Michael with his lance? How are we going to stop people from wanting more and more if not through the power of the Spirit made accessible through religion? And once you have opened up the Pandora's box of the appetites, how are you going to put the genie back into the box? How are you going to be able, with no more than rational arguments, to tell people to use less, to be less covetous, not to be greedy, and so forth? No force in the world today, except religion, has the power to do that.

For the vast majority of people there is no other way to control the great passions within us which have now been fanned by, first of all, the weakening of religion and, secondly, the substitution of another set of values derived from a kind of pseudo-religion whose new gods are such idols as 'development' and 'progress'. But such notions do not have the power to help us control our passions. On the contrary they only fan the fire of those passions. We have been witness during the last generation alone to the ever greater debunking of the traditional religious attitudes towards the world, especially what we call in Arabic *riḍā*, that is contentment with our state of being, a virtue which is the very opposite of the sin of covetousness. Of course, the Muslims have been criticized by the West for a long time for simply being fatalistic in the face of events, for being too content with their lot. This same de-bunking has also been directed towards similar Christian values. But that is because of a deep misunderstanding. Where, in the current educational system in the West, is attention being paid to these traditional virtues? Even from a purely empirical, scientific point of view, these virtues must be seen as being of great value seeing that they have made it possible for human beings to live for hundreds of years in the world without destroying the natural environment as we are currently doing. These traditional virtues that allowed countless generations to live in equilibrium with the world around them were at the same time conceived as ways of perfecting the soul, as steps in the perfection of human existence. These virtues provided the means for living at peace with the environment. They also allowed man to experience what it means to be human and to fulfil his destiny here on earth, which is always bound to trying to inculcate such virtues within oneself.

Another cardinal and central role of religion in the solution of the environmental crisis, one that goes to its very root, is much more difficult to understand within the context of the modern mind set. This role is related to the significance of religious rituals as a means of establishing cosmic harmony. Now, this idea is meaningless in the context of modern thought, where ritual seems to have no relation or correspondence with the nature of reality. In the modern world-view, rituals are at best personal, individual, subjective elements that create happiness in the individual or establish a relationship between him or her and God. That much at least some modern people accept. But how could rites establish cosmic harmony? From the modern scientific point of view such an assertion seems to make no sense at all. But it is not nonsense; it is a very subtle truth that has to be brought out and emphasized. From both the spiritual and the religious perspective, the physical world is related to God by levels of reality which transcend the physical world itself and which con-stitute the various stages of the cosmic hierarchy. It is impossible to have harmony in nature, or harmony of man with nature without this vertical harmony with the higher states of being. Once nature is conceived as being purely material, even if we accept that it was created by God conceived as a clockmaker, this cosmic relationship can no longer be even conceived much less be realized. Once we cut nature off from the immediate principles of nature, which are the psychic and spiritual or angelic levels of reality, then nature has already lost its balance as far as our relation to it is concerned.

Now rituals, from the point of view of religion, are God made. I am not using the term ritual as seen from the secular point of view, as if one were putting on one's gown and going to some commence-ment exercise or some other humanly created action, often called a 'ritual' in everyday discourse today. I am using it in the religious sense. According to all traditional religions, rituals descend from Heaven. A ritual is an enactment or rather re-enactment here on earth of a divine prototype. In the Abrahamic world, that means that rituals have been revealed to the prophets by God and taught by them to man. The 'repetition' of the Last Supper of Christ in the

Eucharist or the daily prayers of Muslims, where do they come from? According to the followers of those religions, they all come from Heaven. In Hinduism and Buddhism one observes the same reality. The differences are of context and world-view, but the fundamentals are the same. There is no Hindu rite which was invented by someone walking along the Ganges who suddenly thought it up. For the Hindus such rites are of divine origin. The Muslim daily prayers, which we have all seen in pictures, were given by the Prophet to Muslims on the basis of instructions received from God. Even the Prophet did not invent them. The Eucharist 're-enacts' the Last Supper which, as the central rite of Christianity, was first celebrated by Christ himself. Now, these rites, by virtue of their re-enactment on earth, link the earth with the higher levels of reality. A rite always links us with the vertical axis of existence, and by virtue of that, links us also with the principles of nature. This truth holds not only for the primal religions, where certain acts are carried out in nature itself—let us say the African religions or the Aboriginal religion of Australia, or the religions of the native American Indians—but also in the Abrahamic world, in the Hindu world and in the Iranian religions. Whether one is using particular natural forms such as a tree or a rock or a cave or something like that, or man-made objects of sacred and liturgical art related to rites carried out inside a church, synagogue, mosque or Hindu temple, it does not make any difference. The same truth is to be found in all these cases. From a metaphysical point of view a ritual always re-establishes balance with the cosmic order.

In the deepest mystical sense, nature is hungry for our prayers, in the sense that we are like a window of the house of nature through which the light and air of the spiritual world penetrate into the natural world. Once that window becomes opaque, the house of nature becomes dark. That is exactly what we are experiencing today. Once we have shut our hearts to God, darkness is spread over the whole of the world. This, of course, is something very difficult to explain to an agnostic mentality. But from a practical, expedient point of view at least it should be taken into consideration even by those who do not take rites seriously, seeing what happened to nature at the hands of those sectors of humanity who no longer perform traditional rites. All religious people who believe in the efficacy of rites and perform them have a way of looking at the natural world and their place in it which is very different from the secularist way that has itself led us to the environmental crisis. You have all read or heard about examples of various religious rituals and their relation to nature, even in lesser known religions. Perhaps the best known, as far as displaying the direct relation between rituals and the natural world is concerned, is the rain-dance of the Native Americans, about which sceptics make jokes. But some people take it very seriously and go to Native American medicine men, the shamans, to try to help them to bring rain. Of course, such a thing is laughed at by official science, but that does not matter for such a science neglects the *sympatheia* which exists between man and cosmic realities.

We have similar rituals all over the Islamic world, the Hindu world, the Buddhist world and in the traditional Christian world. But in the modern Western world it has now become more or less eclipsed, although it has not disappeared completely. In Greece, once you go out of big cities, you still get it and in Italy, in the villages, when there is news of an earthquake, people recite the beginning of the Gospel of John in Latin, which many still know by heart. The faithful recite it in a ritual sense to help recreate balance and harmony with the natural world by calling upon Divine Mercy. I can hardly over-emphasize the significance of this aspect of religion because it is impossible for a human collectivity to live in harmony with nature without this ritualized relationship with the natural world and harmony with God and the higher levels of cosmic hierarchy. If we do not have this relationship, nature is reduced to an 'it', to a pure fact, to a material lump, not in itself, of course, but for us and we must bear all the consequences which such a view entails.

Along with providing a sound basis for ethics, perhaps the most important role of religion in the understanding of the roots of the environmental crisis (and here I would include especially the spiritual element of religion because it is the spiritual, metaphysical and esoteric element of religion which emphasizes this element), is that religion possesses an extensive doctrine about the nature of the world in which we live. That is, religion, when it was integral and not truncated as it has become today in the West, provided not only a doctrine about God, not only a doctrine about the human state, but also a doctrine about the world of nature. And here, by doctrine, I mean knowledge (*doctrina*), not only opinion, but authentic knowledge which is not in any way negated by the scientific knowledge of the world. Every religion provides not only teachings pertaining to the emotional and sentimental realm, not only principles for ethical action, but also knowledge, knowledge in the deepest sense of the term: of God, of the human state, and also of nature. There is no major religion whose integral tradition does not provide such a knowledge. Some religions emphasize one element, some religions another. Certain religions, such as Confucianism, do not speak about cosmogony and eschatology, but they have a vast cosmology. Of other religions, the reverse is true. But these three types of knowledge, that is, knowledge of God or the Ultimate Principle, of the human state and of nature, have to exist in all integral religions.

Now, one does not need to look very far to see what has happened in the modern world. Gradually, from the seventeenth century onwards, first in the West, then spreading in recent decades to other parts of the world, the legitimacy of the religious knowledge of nature has been rejected. Most people who study the views of an Eriugena or a St Thomas Aquinas on nature do so as historians. But their views are not accepted by the mainstream of modern Western society as legitimate knowledge of the world. What has been lost is a way of studying nature religiously, not simply as 'poetry', as this term is used today in a trivializing sense and not of course in a positive one. True poetry possesses a great message as far as nature is concerned, a message which itself is usually religious. In any case modern society has disassociated knowledge of nature from religion as well as sapiential poetry itself, and relegates the religious attitude and knowledge of nature to sentiment or 'simply' to poetic sensibility.

We have wonderful examples of nature poetry in the great poetry produced in the nineteenth century in this country. The Romantic poets produced beautiful poetry about nature. But what effect did it have on the physics departments of the universities? Absolutely none. Precisely because the science that developed in the seventeenth century, through very complicated processes which I cannot go into now, began to exclude from its world-view the possibility of a religious or metaphysical form of knowledge of nature. This science even excluded the poetic view of nature insofar as it claimed any intellectual legitimacy and sought to be more than what some would call 'mere poetry'. Modern science has clung to that monopoly very hard, even in this pluralistic age of ours, in which everything other than science is relativized. Post-modernists usually deconstruct everything except modern science because if this were to be done the whole world-view of modernism along with postmodernism would collapse. So you have a kind of scientific exclusivity and monopoly which has been created and accepted by most although not all people in the modern world. Goethe, the supreme German poet as well as a scientist, rebelled very strongly against this monopolistic claim of modern science. There were also certain scientists, such as Oswald, who was a reputable chemist who rejected scientific mechanism; and one can name others. But these are exceptions to the rule. The rule became that there is no other knowledge of nature except what is called scientific knowledge. And if someone claims that there is a religious knowledge of nature then it is usually claimed that it is based on sentiment, on emotions, or in other words on subjective factors. If, for example, you see a dove flying and you think of the Holy Spirit, that is simply a subjective correlation between your perception of the dove and your own sentiments. There is no objectivity accorded to the reality of nature as perceived though religious knowledge. That is why even symbolism has become subjectivized—it is claimed to be 'merely' psychological, à la Jung. The symbols which traditional man saw in the world of nature as being objective and as being part of the ontological reality of nature, have been all cast aside by this type of mentality which no longer takes the religious knowledge of nature seriously.

During the last thirty years, when the thirst for a more holistic approach to nature made itself felt, something even worse occurred because neither mainstream religion nor modern science showed any interest whatsoever in the religious and symbolic knowledge of nature and the holistic approach to it. The water sought for this thirst seeped under the structures of Western culture and came out in the form of New Age movements, nearly all of which are very much interested in the science of the cosmos. But what they claim is really a New Age pseudo-science of the cosmos. It is not an authentic traditional science because a traditional science of the cosmos always has to be related to a traditional religious structure. In this New Age climate the word 'cosmic' has gained a great deal of currency precisely because of the dearth of an authentic religious knowledge of the cosmos in the present-day world. Somehow the thirst had to be satisfied. So we have had both excavation of the earlier Western esoteric teachings about nature—usually presented in distorted fashion—or borrowings from Oriental religions and their teachings about nature, often distorted. Even the famous and influential book of Frithjof Capra, *The Tao of Physics*, does not really speak of Hindu cosmology or Chinese physics, but only mentions certain comparisons between modern physics and Hindu and Taoist metaphysical ideas.

To be sure there are many profound correlations and concord-ances to be found between certain aspects of biology, astronomy and quantum mechanics on the one hand and Oriental doctrines of nature, of the cosmos, on the other. I would be the last person to doubt that truth. But what has occurred for the most part is not the kind of profound comparison we have in mind, but its parody, a kind of popularized version of a religious knowledge of nature, usually involving some kind of occultism or even some kind of an existing cult. The great interest shown today in Shamanism in America, in the whole phenomena of the Native American tradition (which is one of the great and beautiful primal traditions that still survives, to some extent), with weekend Shamanic sessions, is precisely because such teachings appeal to a kind of mentality that seeks some sort of knowledge of nature of a spiritual and holistic character other than what modern science provides. This phenomenon is one of the paradoxes of our day which has not helped the environmental crisis in any appreciable way. Indeed, it has created a certain confusion in the domain of religion and created a breach between the mainstream religious organizations which still survive in the West—whether they be Catholic, Protestant or Orthodox—and these pseudo-movements and the New Age phenomenon which they rightly oppose. The fact that these pseudo-religious movements are very pro-environment, yet in an ineffectual manner, has caused many people in the mainstream to take a stand against the very positions which they should be defending. So we have the paradoxical situation in America today where the most conservative Christian groups are those which are least interested in the environment. This phenomenon was not originally caused by the rise of the New Age religions but is certainly related to it and strengthened by it.

It is now necessary to say something about what constitutes the religious view of nature. It is important for us to understand that ordinary human beings in all civilizations deal with religion to the extent that it concerns their everyday life. Only the sages, the seers and the saints are able to reach the heart of religion, but they always bring something from the heart and disseminate it throughout the whole body of the religious community. Now in the Western world (the Christian world), and the Islamic world which has a religion which is a cousin of Christianity, and Judaism (Judaism, of course, sharing something of both worlds), the religious science of nature was not part and parcel of the everyday teaching of religion as far as ethics were concerned. On the everyday level these religions teach that one should be good, one should love one's neighbour, which includes beings other than man, one should give tithe and alms, one should say one's prayers, etc. But the fact is that in the traditional ambience there was also always present this inner spiritual message and knowledge concerning nature which was accessible to those in these civilizations who sought such a knowledge. It was this message that first became eclipsed in the West before the onslaught of a secular science took hold of the predominant world-view. It was not that first a secular science took hold in the West and then the spiritual understanding of nature was eclipsed, but the other way around. The loss of the Christian sapiential view of the cosmos during the Renaissance left a vacuum which after two centuries came to be filled by a secular science which could not see in nature anything other than quantity and motion, as Galileo asserted.

One can now ask what is this religious view of nature? The religious view, as understood by the ordinary sense of the believer, is that this world is created and sustained by God; it is His creation; or one could in the non-Abrahamic context say that it is the manifestation of the Principle, of let us say the Tao of the Far Eastern traditions. The Origin does not have to be identified in name as the Abrahamic God, but it is always the Divine Principle which is none other than God. On the deeper level the religions teach that there is something spiritual and ultimately meaningful in nature. On this deeper level it is this meaning which has to be deciphered and understood. This deeper meaning is not, however, like the exoteric teachings of the religion which are meant for one and all. We are talking here about a type of knowledge that is not accessible to everyone, but only to those few capable of metaphysical intuition and who are willing to master the language in which the Divine Message is written upon the leaves of the cosmic book. Knowledge is not evenly and democratically distributed in any domain. You only have a few great mathematicians in Great Britain and the other fifty-four million people have only a rudimentary grasp of mathematics. It is like that in every field. With metaphysics it is the same; there are only a few who can grasp its teachings: here it is certainly not quantity that is important.

There is a fine book by William Chittick on the cosmology of Ibn 'Arabi which is called *The Self-Disclosure of God*. It is the most extensive book on Sufi cosmology in English, one which brings out clearly this esoteric and metaphysical knowledge of nature in its Sufi context.[113] This type of knowledge was always reserved for the few but is now made available publicly to all those endowed with the necessary qualifications to understand such doctrines. God self-discloses Himself. It is not only a question of creation of the world by God, but there is also creation in God. Creation first takes place in divinis and then is manifested externally. Therefore, although creation is not God, it is not completely divorced from its Divine Origin. We might say the cat is also God's creation, and so, at best, we should not molest the cat in the street. The truth, however, is much more profound than that. The cat is ultimately a manifestation of some aspect of the Divine Reality itself. Therefore the essential reality of the cat is in the Divine Reality in the ultimate sense. All of nature is the self-disclosure of God Himself to Himself, without which it would be literally nothing.

I do not want to go into these complicated philosophical concepts of archetypes, Platonic ideas and so on which we find in different languages in all of the traditions. I am speaking in simple language for the purposes of this discourse. Put simply one can say that everything in the world is a divine presence and witness to God. The Qur'ān is very specific about this: 'The seven heavens and the earth and all that is therein praise Him, and there is not a thing but hymneth His praise'.[114] So every time we destroy a species, we are destroying a prayerful being. It is like murdering someone while he is praying. It is as abominable as that. In the time I am delivering this lecture we will have destroyed several species. It is amazing what we are doing. They say that 30 per cent perhaps of all the species in the world will have been destroyed in the next twenty years. This horrendous fact is directly the result of a type of knowledge of nature having been lost, a knowledge which is based on the nexus between all creatures and their Divine Source. If I make such a statement, one might think that I am talking as a dreamy Persian philosopher or poet, and what I say has nothing to do with real knowledge. Knowledge of the cat is the size of its tail, the structure of its bones and so forth and so on. To be sure that is all part of the knowledge of the cat. But the knowledge of the cat is not exhausted by our knowledge of its physical aspect. What we learn about the physical aspect of the cat is in itself remarkable. The structure of the cat, as with any other creature, is amazing enough, consisting, as it does, of the smallest units of life, which have not evolved from simplicity to complexity but appeared on the natural scene with remarkable complexity. But the religious view neither limits itself to this external knowledge nor does it deny such knowledge. It deals with the very being of a creature as locus of Divine Presence. The result of such knowledge is that we must live with the creatures of the world, not only by necessity, but also because of our own spiritual welfare. The destruction of nature is ultimately the destruction of our own inner being and finally our external life as well. Of course, from the point of view of cause and effect, it is the reverse; it is the pollution of our inner being that has caused the pollution of the natural environment. It is our inner darkness that has now extended outward into the world of nature. The chaos of the outward reflects like a mirror what has happened within ourselves.

Without resuscitation of this religious and metaphysical view of nature, everything else we say about the environmental crisis is just cosmetics and politics. We have to experience the profound rebirth of our conception of the world as *temenos*, a sacred precinct, as the Greek word signifies. It is not only the mosque, the temple or the church that is a sacred precinct; it is not only human beings, but the whole world, everything in creation that bears testimony to the Divine Presence which is therefore *temenos*. As this verse of the Qur'ān, which I cited for you, asserts, it is not only human beings who sing the praise of God. Everything has its own tongue with which it praises Him. Everything sings the praise of God by virtue of its very existence. We might not understand the language, but the song of praise is there. It is very significant that so many of the great saints of the Abrahamic world who were sensitive to nature claimed to have some kind of communion with nature beyond human language. I mean such figures as St Francis of Assisi who spoke to the birds; we have examples of that in Islam and in fact in all religions, not only the Abrahamic ones. The religious view of nature requires of us a complete re-understanding of what nature is, and who we are as human beings who act upon nature, because it is impossible to discuss nature without discussing the image that we have of ourselves.

The fact that in our religions it has been told to us that all of nature praises God presupposes that God wants us to know this reality, that He has bestowed us with the means to understand such a message. That itself is enough to prove to those with eyes to see and ears to hear that in the deepest sense we are God's vicegerents here on earth. We have a viceregal and pontifical function in this world. Of course, many of us have now abdicated that function. Modern man has not only destroyed most of the monarchies in the world, but has also tried to destroy the king or the queen within. There is a regal function which the human being has, what I have called a pontifical function, the function of being a bridge between Heaven and earth. The word pontifex originally meant 'bridge-builder'. We all have this function, the pontifical function, or the regal function within us which we have now cast aside. The religious point of view means that our relation to nature, the very fact that we accept that nature is the locus of God's Presence, imposes upon us the necessity to fulfil that viceregal function.

Now God is not only the creator of nature but also its sustainer and nourisher. To fulfil that caliphal function (khalīfah or vicegerent is an Islamic term; I could equally use terms drawn from other religions) therefore means not only that we must be able to make use of nature but also that we are required to sustain and protect nature. The idea that nature is out there only for our use is not only un-Islamic, it is also not a traditional Christian idea. It has been foisted upon Christianity by certain later interpretations and also by certain modern critics of the environmental crisis trying to blame Christianity (and also Judaism) for the crisis which is much more the consequence of the secularization of nature and what I call the absolutization of the human state, beginning with Renaissance humanism and all that came after it. In the religious perspective, even religions which accord a very special position to human beings, the human being is never absolutized. For example, in Christianity people have certain rights, but the rights of God come before every-thing else.

Ours is the only time in history in which human beings claim for themselves absolute human rights with disrespect for the rights of either God or the rest of His creation. There is nothing that is more dangerous for the environment than this view of human rights. It is going to kill us all. If we continue our foolish ways, our great grand-children will have only one right and that is to die. Meanwhile, in the name of the rights of human beings, we are destroying the web of life in the planet. This absolutization of human rights and claims is one of the most dangerous ideas that has come out of civilization, no matter what the newspapers and politicians say. Its dangers are much greater than its benefits, albeit there are certain benefits. There is hardly anything that is completely black or white in history. Some-thing completely black would not exist; something completely white would be celestial. Everything in this world is a *mélange* of the two. I am not saying there are no contingent benefits to this way of conceiving of our rights here on earth. But from the point of view of the natural environment, what I call the absolutization of the human state, that is taking God's rights and nature's rights away and giving them all to man, is at the very heart of the environmental crisis. It is this dangerous possibility that the religious view of man and nature always avoided. No matter what the Qur'ān or the Bible says about man's right to make use of the world of nature, they also emphasize man's responsibility towards nature. The existing situ-ation, in which we always shout about rights without mentioning our responsibilities, is a trait of modernism. Ours is the first period in human history in which irresponsible people believe that they have absolute rights with which they are born. People are brought up to believe that even if they do not fulfil their responsibility, they have rights. This certainly is not the traditional point of view.

If one could resuscitate the religious view of nature and of man, one would be able once again to relativize our importance and our so-called needs, which is the only way that would allow us to survive. What is absolute in the human being is not his animal needs; it is his spiritual essence. It is the manifestation of the Absolute Itself in each being. What we have done in the modern world is to transform that spiritual reality into a purely earthly one, thereby endangering the earth. I must repeat here what I have said before, that one could understand the whole tragedy of the environmental crisis by realizing that modern man has taken the notion of God as Infinite and horizontalized it, trying to realize the Infinite in a finite world, that is, substituting ever-changing things, ever-changing objects, going from one thing to another, to satiate her or his thirst for the Infinite in the physical domain which is by definition finite. But our thirst for the Infinite can only be satisfied vertically. The fact that we are never satisfied with earthly needs derives from a profound metaphysical truth, from the truth that our soul was not made for the finite. We are made for the Infinite. But the fact that this thirst is now horizontalized and is turned to never-ending material needs causes us to be always unhappy about what we have. That we continue to dream that the next object we acquire is going to make us happier is precisely the result of forgetting who we are, that essential nature of the human state emphasized by all religions.

The religious view of man and nature which has been lost, a loss which is the root cause of the environmental crisis, is also based on a very important principle which may be difficult for us to grasp but whose loss is, I think, also one of the root causes of the separation of man from nature. That principle is the idea of law or order in the religious sense in the world of nature, an idea which is to be found in one way or another in all the sacred traditions. From Tao to *dharma*, to *ṛta* to *Sharī'ah*, to *nomos*, whatever term is used to designate this reality in different traditions, whatever the difference of emphasis, none the less the essential content of these terms remains essentially the same. They all demonstrate that there is an order that governs man as well as nature, from which comes our modern word *cosmos*.

The Greek word *cosmos* means both order and beauty, which is extremely significant; the idea of order as beauty. All of these cardinal concepts possess a religious as well as a cosmic sense.

For example, *nomos* in Greek meant not only the laws by which the planets move, but also the laws which govern human life, laws by which the wise man should live. In Islam the Greek word has come into Arabic as *nāmūs*, and we use the term *nāmūs* as being equivalent to the *Sharī'ah*, the Divine Law (which is a Qur'ānic term), but also we use it for laws of nature. The very word *sunnah* in Arabic, which means both tradition and the mores of the Prophet, is used in the Qur'ān also as *sunnat Allāh*. We have not only the *sunnah* of the Prophet, but also the *sunnah* of God, which is precisely the laws and principles governing the existence of all creatures. *Sunnat Allāh* is ultimately the laws and norms which govern religion as well as God's creation, the principles by which the world functions, what we observe as the laws of nature. The same holds true for *dharma* even if this term is not associated with the personal God of monotheism. Nearly all contemporary Buddhist thinking about the environment rotates around this single concept of *dharma*. The word *dharma* (translated sometimes as duty, law or principle) is at the fore of discussions in this matter because *dharma* is not only related to the correct way of living, but also to the principles according to which things are what they are. In fact everything in turn has its *dharma*. The streams, the flowers and the mountains have their own *dharma*; that is why this term is so difficult to translate into English. The same holds true of the Hindu term *ṛta*, which is not only the law for human beings but also for the cosmos. The religious world-view points to a kind of mystery—because it is really a mystery from a purely human point of view—the mystery of the relationship between laws that should govern us morally and spiritually and the laws that govern the universe.

There is a profound relation between the two. There are currently some attempts by a number of scientists to try and rebuild this bridge from the other side. Professor Wilson, the famous evolutionary biologist from Harvard, has just published two essays which have been the cause of much discussion in the American intellectual establishment. He begins by saying that the humanities and science should come together and overcome the separation that now exists between them. He further proposes that they should do so by developing the humanities on the basis of biology. He proposes that one should develop ethical and social laws for society on the basis of what people like him have discovered in the biological world. This is not what the religious view has in mind at all, because none of us wants to live under one form or another of social Darwinism, applying what people wrongly call the 'laws of the jungle' or some other so-called biological law to human society. In fact the image we have of the 'law of the jungle' is itself a prejudice, because if it were the only law involved all the animals would already have eaten each other. In truth we find that an incredible harmony pervades the jungle and the relation between living and non-living beings, a harmony to which little attention is paid by many scientists.

This kind of idea of law pertaining to both society and the cosmos is not what I am talking about. Rather, I am saying that traditional man believed that his way of living—whether a particular individual chose to follow God's laws (or the Tao, or the Principles of things— you can be non-theistic and talk about these terms as well)—was related to how the world functioned. This principle was the basis of the hieratic function of the priest-kings in various traditional civilizations. For example, the Chinese emperor was the bridge between Heaven and earth and performed certain rituals which were related to the harmony of the cosmos. The same principle can be observed in the function of the pharaohs of ancient Egypt, Melchizedek in the Hebrew tradition, Saoshyant in Zoroastrianism, Kalki in Hinduism and many, many others. We can also see examples of it in historical times in the three monotheistic religions.

Now it is not possible to recreate those institutions until the return of Melchizedek or the Mahdi, but it is possible to bring back the idea that there must be some kind of relationship between our ethical norms and the way we deal with the world of nature. We cannot have the world of nature as simply a pure 'it', an object totally bereft of value in the classical Galilean and Cartesian sense of the term, with an island populated by ethical beings with ethical concerns which are irrelevant to the rest of creation. The whole debate that is now going on about the question of animal rights versus human rights, how we should deal with the animal world, these are very complex questions and related to this link between ethical laws and cosmic laws. Thank God such debates have come to the fore at this late hour. If we had thought about them when we were beginning to destroy all the forests in Africa and America a few centuries ago, we would not probably be in such a dire situation. In modern times nobody even thought of these matters until now when we are made aware that if we continue in our present course, having killed off all the big animals we will be left with only a few small creatures. Our neighbours will be just a few species of small creatures which we cannot destroy so quickly.

My time is running out but I want, in the last part of this lecture, to say a few words about the complicated process by which the loss of the religious view of the environment took place in different parts of the world and especially in Western Christianity; not the whole of Christianity, for Greek Orthodox and Russian Orthodox Christianity have pursued a different path, at least until recently. Also in this matter other religions and civilizations of the world have followed other historical lines of development which are very different from that of the modern West. Although the fruits of modern science and technology in the form of Boeing jets go to all of the continents of the world, and the political heirs of Mahatma Gandhi have atomic bombs, in other religions and civilizations there did not occur this long process whereby gradually the world of nature was secularized with the effect that a respected thinker from the modern West could no longer think in religious terms about the world of nature. The process which took place in the West did not occur in other parts of the world, even in Eastern Christianity and Oriental Judaism. The question of distinguishing between various views of nature is not, therefore, just one of religion but also of geography, of a particular development of religion and type of culture and civilization that came into being in Western—mostly North Western—Europe, in modern times and then spread to America and elsewhere. In order to understand the present environmental crisis, it is important to grasp this point, especially if one is looking seriously for a solution.

Non-Western people in general do not understand the process by which the secularization of nature took place in the West. They rarely grasp the significance of the fact that modern science is not simply an 'objective' knowledge of nature but is based on a particular philosophy and also on the idea of domination over nature, especially in physics and chemistry where one aims at a controlled situation in the laboratory in order to analyse a part of the physical world under the complete control of the experimenter and thereby gains the means to better dominate the physical world. This pursuit of controlled power has succeeded and thereby allowed Western science, as applied in various technologies, to gain remarkable domination over nature during the last two hundred years. It is this 'method' along with certain philosophical presumptions concerning the nature of reality that has landed us in our current predicament. Although not aware of the philosophical background of the rise of modern science and the idea of domination over a segmented nature 'coerced' into situations of controlled experiment, non-Western people are none the less fully aware of the relation between the applications of modern Western science and power. They also think that this science can help them gain power and domination over their own affairs, but without thinking of its consequences whether they be ethical, spiritual or environmental. That is why, in the non-Western world, all forms of government, from the left to the right, from the religious to the anti-religious, subscribe to the applications of modern science in the form of modern technology and espouse the cause of industrialization with as great a rapidity as possible. This fact in itself is quite amazing given the survival of the religious view of nature among their people. I give you the following example, which is annoying but true.

For several years in the 1970s I was the president of Iran's most important scientific and technological university. The reason I accepted this task was to be able to create intellectual and cultural responses which could protect Persian culture in the face of the powerful onslaught of Western science and technology. Now, our university was building a plant for nuclear energy in the port of Bushehr in the Persian Gulf. The students in the university who were opposed to this project would come out nearly every day with pronouncements of how terrible it was to carry out this project and were turning the issue into a political one against the government. I was happy to agree on this issue and said at the time to the authorities that in this question the students are right and I tried several times to stop the plant's development, as I believed that from an environmental point of view it was dangerous and not really needed. But my voice was not heard and the plan went ahead. As soon as the Islamic Revolution of 1979 took place, the further building of the plant was stopped. Nineteen years later, however, with the expense of several billion extra dollars, the plant is now being completed. It is a telling fact that, whether one has the royal regime or the Islamic Republic in Iran, the monarchy of Saudi Arabia or the secular Ba'th party of Iraq, the Hindu Nationalist Party (the BJP), or Communist China, or the very different political systems of Malaysia or Indonesia, no matter where one looks, one sees that the attitude towards modern Western science and technology is in each case nearly the same. The reason is the misunderstanding of non-Western people of what is really involved; of the dangers which threaten their religion and also endanger the whole earth, of the impossibility of repeating the errors of the industrialized West in every corner of the globe in the name of gaining independence from the domination of the West. This is the reason why interest in the whole environmental issue has started so late in the non-Western world.

In the West, however, one has had a very different process. Gradually, step by step, the religious view of nature was lost and the mechanistic point of view replaced it. And now, after three or four hundred years (really since the trial of Galileo), the religious establishment in the West is trying, one way or another, to reformulate a theology of nature. For that very reason I think that the Western thinkers who are dealing with this issue have a very grave responsibility, not only for the Christian or the Jewish world, but globally. Quite obviously they, having gone through all of these battles, are much more aware of all the issues involved than many people in the non-Western world who are only now turning to this question. From the other side, the thinkers of other religions have the advantage that amidst their co-religionists the sense of the sacred in nature and the legitimacy of a religious knowledge of nature has not been lost to the degree one sees in the West.

Let me conclude by giving a few practical suggestions. What can be done at this late hour to try to reverse the critical environmental situation? I am certainly not opposed to individual or group efforts to try to clean up the Thames, or to prevent a particular tree from being cut; thank God for that. Such actions can only delay but cannot prevent our mass suicide. The fact that we are murdering creation is what has to stop. The only action for the preservation of the environment which is likely to be effective must be based first of all on the thesis that we are responsible for our actions: we cannot sit down and do nothing with the pretext that this has been destined by God or is inevitable because of the march of modern technology. God holds us responsible for what we do and what we do not do but could and should do. When we accept and acknowledge the freedom given to us to not only destroy nature but also to live in harmony with it and then ponder freely all the possibilities this may give rise to, we see that our only possible action is not even action in the ordinary sense but a change in our state of being and consciousness.

There is no other way than to change our whole world-view. That means, bluntly, that we are faced with the stark choice of either the death of the modern world or the death of humanity. There is no third choice. By the death of the modern world I do not mean the death of all human beings living in the modern world. I mean the death of the point of view which we call modernism, which is based on the severing of the relationship between man and the Divine and between man and nature as spiritual reality. This severance has to be repaired, which means that the current modern world-view must be discarded. There is no other way. All compromises at this stage of history are the worst kind of treason. It is much better not to compromise at a time when compromise simply destroys the opportunity for us to do something serious. We have had too many compromises with the truth in the modern world. I always say jokingly that the modern world is characterized by somebody getting up in the morning and saying two and two is six. Then they debate all day long and some nice, liberal pacifist comes along and says, do not argue, do not fight, we will make it five and settle the count. The world has gone on day after day, year after year in this manner and can no longer afford to do so. When we come to the question of the environment, I believe that like any other basic matter involving truth and falsehood one has to be categorical and no longer compromise in any way any principle that is involved. I do not see how the modern world, with its presumptions, can survive. No more can humanity survive while holding to a world-view which is false in its very foundations. How can we keep the word development as currently understood in our vocabulary, and elect people who believe in continuous material development, without committing suicide? I would be glad to be proven wrong. I am not only a philosopher, but have also studied the sciences and if somebody can show me on the basis of serious scientific evidence that my assessment of the environmental crisis is wrong, then I shall accept and thank God for it. But I do not see that if you extrapolate all the present trends, as scientists tend to do all the time, how it will be possible in the future for the earth to sustain human life, not to speak of life with quality.

It is in light of this situation that the religious view of nature becomes so significant. The resuscitation of the religious view of nature implies of course a very radical step. First of all it implies that religion must challenge not what science says within its own legitimate domain but the monopolistic claims of science. Science is a view of the natural world, and if its applications in the form of technology had not become wedded to greed, had it not caused such over-population, etc., it would have been an exceptional achievement, which in a sense it is despite its philosophical shortcomings and negative consequences. But that is another question. As far as the natural environment is concerned, however, there must be a space within the mind of modern man for a view of nature other than the modern scientific one.

There must also exist profound criticisms of all views which would try to negate and gloss over the traditional understanding of the relationship between man and the world. We must overcome this hypnotic trance which causes us to make all kinds of false assertions and deny age-old truths by claiming that we are now in the space age or something like that, as if there were not still beautiful donkeys in Cyprus and therefore logically we could call our age the donkey age. What is it that makes the metaphor of the space age so important for us at the expense of so much else that still exists in our world? We must ask this question and wake up from this trance. Our problem is not solved by talking of the space age. Our problem is what we are doing here on earth, to ourselves, to our families, to our greater family of living creatures, to the non-living creatures of the earth and to the skies that we are also polluting.

There must be a re-assertion of the religious view of the world without compromise, without being intellectually embarrassed, as one sees so often among Western theologians. There is nothing intellectually embarrassing about the religious view of nature if one were only to understand it in depth. If one were to understand the metaphysics that stands behind it, one would realize that it is in fact based on intellectual knowledge of the highest order. My dear friend, Dr Martin Lings, apparently spoke about metaphysics and the perennial philosophy to some of you a couple of weeks ago. That kind of metaphysics which lies at the heart of all authentic religions, if fully understood, is not something embarrassing, to say the least. One can present it to the most acutely aware and rational minds among the philosophers or scientists of our day without the least amount of inferiority complex.

It should be remembered, however, that one can never build something unless one has cleared the ground. That is, there is a need first of all for an in-depth criticism of all the errors of the modern world in the light of the truth of traditional teachings. Often I and people like me are told why do you keep criticizing things, why can you not just state your ideas. But it is impossible to be indifferent in the face of error. Knowledge always implies both truth and false-hood. We live in a world in which outside of science the word truth is very unfashionable. Few in official academic life talk about the truth any more; it is the last word one talks about in fashionable intellectual circles in the West today. Consequently, one also does not need to speak of falsehood, which in the domain of religion has to do with heresy (a term that few even in the churches talk about or use; today's culture prefers to speak of 'alternative life style'). Truth is no longer a significant category. But from the point of authentic knowledge, knowledge cannot exist without truth. In physics, chemistry or biology, if someone comes up with a new theory about something we cannot say, all right, you keep your view and I shall keep my view and we will be friends. This is because from the point of view of science you have to test things out and either one theory or the other is true as the term truth is defined in the particular science in question. In the same way when we come to the sapiential dimension of religion, we cannot remain impervious, even in the name of charity, not to say indifference or intellectual laziness, to what negates the truth, and we must be critical of any world-view which would do violence to our nature, to our destiny, to our relationship to the world, to the animals and the plants, and ultimately, of course, to God. The ultimate question for us, the ultimate challenge, is: who are we? What are we doing here? And the response has always been that we are here first of all to remember who we are; we are here to remember what the world is in its spiritual reality; and above all we are here to remember God Who is the source of both the world and ourselves. Only through this remembrance can we regain the vision necessary to live at peace with God, with ourselves and with His creation, with all of His creation both animate and inanimate, that by His Mercy sustains and nourishes us even if in our ignorance we are unworthy of all His blessings. The attainment of this peace alone can ameliorate the critical condition of the world about us by establishing order again within ourselves and by opening our eyes to the vision of the natural world as the theatre of His endless theophanies.[115]

EDWARD GOLDSMITH

The Cosmic in Art, Architecture and Ecology at the Millennium

🖋

EDITORS' NOTE: *Edward Goldsmith's contribution to the series was largely inspired by a paper which had already appeared elsewhere; we are grateful for the permission to reprint it here.[116] Towards the beginning of his remarks, he related an anecdote which captured the spirit of much of his lecture, and also of the essay given here.*

I once took a taxicab in Wellington, New Zealand and the driver happened to be Samoan. I asked him if he owned the cab and he replied, 'No, I can't, because I am Samoan'. I said that I did not know there was a law in New Zealand forbidding Samoans to buy taxicabs. He answered that this was not the point, the point was that 'in my society, if we make any money we must distribute it amongst our family and community members, so if I do that how can I put aside the money to buy a taxi? But I am thinking of becoming like you a *Pakeha*', (which means a white man) 'and then I am going to tell my family and community to go to hell, and I will put aside the money to buy a taxi'.

*

The Millennium coincides with a widespread yearning for individual and Earth healing. Individuals and societies, global and local, and the whole Earth community suffer as never before under unsustainable human impact. The healing ministry should be broadened to include the earth, the living soil, plants, animals, water and climate, and the science and technology, which, when arrogantly misused, threaten the very continuation of our species and the biosphere as we know it.

Edward Echlin[117]

Contrary to what mainstream scientists tell us, I have consistently argued that natural systems at different levels of organization seek, consciously or not, to maintain the order of the larger wholes of which they are part. The biologist Ludwig von Bertalanffy was struck by the 'whole maintaining character' of life processes at the level of the biological organism.[118] So was the Austrian biologist Ungerer, who was so impressed by this process that he decided to replace the biological consideration of purpose with that of wholeness.[119]

That the constituent parts of any natural system must strive to maintain its overall order is clear, because those parts have evolved to fulfil their specific functions within that system, and are thereby totally dependent on its preservation for their welfare and indeed for their survival. Eugene Odum, whose *Fundamentals of Ecology* was the standard textbook in American universities for decades, points out that 'the individual cannot survive for long without its population, any more than the organ would be able to survive for long as a self-perpetuating unit without its organism'.[120] Thus children brought up in a broken home, as any social worker will confirm, will often tend to be emotionally unstable and have a far greater chance of becoming social misfits, delinquents and criminals.

The family, however, cannot thrive as a little oasis of order in a sea of social disorder. It needs to be part of a cohesive community, which is of such importance in the traditional world that people cannot imagine life outside of it. Nor, of course, can individuals, families and communities survive if the order of the natural world or of the ecosphere is destroyed, as even the most extreme adept of the cult of selfishness will soon realize.

Unfortunately, this key principle only becomes apparent when life processes are seen in terms of their relationships with the whole of which they are part. Mainstream scientists who insist on looking at life processes in isolation from the whole—the whole whose very existence most of them choose to ignore—continue to see those processes as random, malleable, goal-less, and self-serving. This interpretation could not be better illustrated than by the writings of Professor Richard Dawkins of Oxford University, for whom there is no selective advantage in displaying any concern for the stability and integrity of the larger whole.

If behaviour is looked at reductively, there is no way in which its 'whole-maintaining' function can be established, and hence no way of distinguishing between behaviour that serves to maintain and that which serves to disrupt the order of the living world. This key distinction is foreign to mainstream science, though critical to early archaic religions. A brief consideration of some early architecture and weaving will demonstrate some of these principles of wholeness.

The anthropologist Henrick Kraemer notes how, in primal societies,

> the dominating interest is to preserve and perpetuate social harmony, stability and welfare. Religious cults and magic practices have chiefly this purpose in view. Everyone who has lived with a 'primitive people' and has tried to immerse his or her mind in theirs, knows the deep-rooted dread fostered towards any disturbance of the universal and social harmony and equilibrium. Whether a violation of this harmony issues from the universal sphere—for example, by an unusual occur-rence in nature—or from the social, by a transgression of tradition or a disturbing event, it calls forth a corporate and strenuous religious activity towards restoring the harmony and thereby saving the fertility of their fields, their health, the security of their families, the stability and welfare of their tribe from becoming endangered.[121]

Most practices of primal peoples are geared to the achievement of this same end, whether these be their agricultural activities; the technologies they use; the designs of their houses, their temples and their settlements; or the performance of sacred rituals. Beyond their utilitarian functions, these practices all serve to maintain the order of the cosmos. It is worth noting how totally irreconcilable this approach is with the principle that all living things are fundamentally egoistic, individualistic and aggressive, a principle underlying Neo-Darwinism, sociobiology and (I am afraid to say) modern reductionistic ecology.

For primal people, a plan to build a new village or city meant first building a holy house or temple, on the cosmic model. The settlement that subsequently surrounded that structure was thus integrated into the cosmic hierarchy, or cosmicized. The traditional ceremony performed for that purpose was, according to Eliade's description, a re-enactment of the original act of creation, or cosmogenesis.

Reichel-Dolmatoff demonstrates how the temples built by the Kogi Indians of the Sierra Nevada of Santa Marta are still seen as small-scale versions of the cosmos:

> Kogi temples are meant to be models of man's relationship with the cosmos, models that convey a sense of world order and, simultaneously, are interpreted as the body of the Mother. Each post, beam or rafter, up to the smallest detail of roof construction, thatch or vines, used in tying together the different parts, has its specific symbolic values. A temple construction can be read as an anatomical model, a geographical model, a model of social structure and organization, or priestly ritual, or of the upper and nether worlds; it also is an instrument for astronomical observation.[122]

Coomaraswamy tells us that 'man has always . . . correlated his own constructions with cosmic or simple supramundane prototypes. For example, the Indian seven-storeyed palace (prāsāda) with its various floors or "earths" (bhūmi) has always been thought of as analogous to the universe of seven worlds'. Coomaraswamy quotes

Mus in his great monograph on Barabadur. Mus tells us that the Buddhist stupa cannot be understood simply from the 'functional point of view'. Its importance resides in its symbolic meaning. For Mus, the stupa 'represents a universe *in parvo*', the axis of the stupa representing the axis of the universe, the dome representing the heavens.[123]

This aspect of symbolic representation was also true of the ancient Temple of Jerusalem. According to the *Midrash Tanhuma*, the Temple 'corresponds to the whole world and to the creation of man who is a small world'.[124] In an ancient Jewish legend, Yahweh orders Moses to build him the Tabernacle. 'But how shall I know how to make it?' Moses asks. Yahweh answers:

> Do not get frightened . . . just as I created the world and your body, even so will you make the Tabernacle . . . . You find in the Tabernacle that the beams were fixed into the sockets, and in the body the ribs are fixed into the vertebra, and so in the world the mountains are fixed into the fundaments of the earth . . . . In the Tabernacle there were bolts in the beams to keep them upright, and in the body limbs and sinews are drawn to keep man upright, and in the world trees and grass are drawn in the earth. In the Tabernacle there were hangings to cover its top and both its sides, and in the body the skin of man covers his limbs, and his ribs on both his sides, and in the world the heavens cover the earth on both its sides. In the Tabernacle the veil divided between the Holy Place and the Holy of Holies, and in the body the diaphragm divides the heart from the stomach, and in the world it is the firmament which divides between the upper waters and the lower waters.[125]

By seeing his body, his house and his settlement as reflecting the same critical order—which is also that of his society, of the natural world and of the cosmos itself—vernacular man understands that his life is subject to the same single law that governs the cosmic hierarchy, and that he is a participant in the great Gaian enterprise, whose goal is to maintain the critical order of the cosmos.

Titus Burckhardt shows that early Christian churches were also once designed on the cosmic model. He makes this quite clear in his beautiful book on Chartres Cathedral, wherein he describes the body of Christ as 'inscribed in the ground plan of the church'. The cross itself was 'formed by the axes of the heavens'; Christ's head 'lies towards the east, His feet towards the west, His arms and hands extend from north to south'. According to the church fathers Jerome and Basil, 'the axial cross of the heavens is the pre-ordained prototype of the wood on which the Saviour was nailed'. Indeed, for the people of antiquity the cross represented the axes of the heavens and was 'the direct expression of cosmic law'. It was assumed that if a transgressor of this law was executed on a cross, 'this was in order to re-establish, both symbolically and practically, the disturbed cosmic equilibrium'.[126]

The cosmic symbolism of early Christian churches could not be clearer than in the case of the cathedral of Edessa (now called Urfa) in northwest Mesopotamia, which was once one of the greatest centres of Christendom. The cathedral was built in the sixth century, like Hagia Sophia in Constantinople, and like the latter was dedicated to Holy Wisdom. The description provided by an early Syriac hymn could not be more illustrative:

> Wonderful it is that this building in its smallness resembles the wide world, not through its size, but in its character: water surrounds it, just as the ocean surrounds the world; its roof is wide like heaven, without pillars, vaulted and everywhere closed, and decorated with golden mosaics as is the firmament with shining stars.
>
> Its noble cupola resembles the heaven of heavens. The upper part of the building rests on the lower part like a helmet. Its wide and splendid arches represent the four sides of the world. Through their multiplicity, its colours recall magnificent rainbows.[127]

Traditional societies did not produce artefacts designed to satisfy none but utilitarian purposes. All products were replete with symbolic meaning, based on knowledge that had a superhuman origin. This was true of basketwork, of pottery, and in particular of weaving. Thus the weaving of the cosmic veil that in the original Temple of Jerusalem separated the *hekal* (representing the earth) from the *debir* or Holy of Holies (representing the heavens) had to be made in accordance with very strict and ancient procedures.[128]

The veil itself, as Josephus tells us, represented the created world; it was

> ... of Babylonian tapestry, with embroidery of blue and fine linen, of scarlet also and purple, wrought with marvellous skill. Nor was this mixture of materials without its mystic meaning: it typified the universe. For the scarlet seemed emblematical of fire, the fine linen of the earth, the blue of the air, and the purple of the seas; the comparison in two cases being suggested by their colour, and in that of the fine linen and the purple by their origin, as the one is produced by the earth and the other by the sea. On this tapestry was portrayed a panorama of the heavens ....."[129]

The ceremonial tunic of the high priest of the Temple of Jerusalem was also replete with cosmic symbolism, as is pointed out both by Josephus and Philo of Alexandria. Josephus tells us that the tunic

> signifies the earth, being of linen, and its blue the arch of heaven, while it recalls the lightnings by its pomegranates, the thunder by the sound of its bells. His upper garment, too, denotes universal nature, which it pleased God to make of four elements; being further interwoven with gold in token, I imagine, of the all-pervading sunlight.[130]

Not only the clothes woven for high priests or the fabrics of the veils and curtains of the temples in which they officiated were of cosmic design. In traditional societies, cosmic design affected weaving in general. One obvious example is the weavings made by the Kogi Indians, fabrics filled with symbolic meanings. The Colombian anthropologist Geraldo Reichel-Dolmatoff, who spent some forty years studying the Kogi and other Colombian Indian tribes, tells us that when a Kogi starts to weave a piece of cloth he sings the following:

I shall weave the fabric of my life;
I shall weave it white as a cloud;
I shall weave some black into it;
I shall weave dark maize stalks into it;
I shall weave maize stalks into the white cloth;
Thus I shall obey divine Law.

These are not empty words, as Reichel-Dolmatoff notes: 'The dress a Kogi wears is a "fabric of life" and by weaving a piece of cloth a man is "weaving his life". He is symbolically organizing his personal feelings and his social existence by the act of weaving.'

For a Kogi, 'a man's thoughts are like threads: cotton thread is white, "good" thought, and the act of spinning represents the act of thinking'. The Kogi explain: 'To spin is to think. When one sits and twists the thread on one's thigh, one thinks a lot: one thinks about one's work, one's family, one's neighbours, everything. The yarn we spin is our thoughts.'

More significantly, during the act of weaving these thoughts are interlaced into 'the web that is society, the weaver's social relationships, his social network'.

The critical role played by weaving in the social world of the Kogi is quite consistent with Kogi cosmogony, for the spindle represents the centre of the Earth—the *axis mundi* that holds up the heavens. When the Mother Goddess created the earth, she pushed a spindle 'upright into the newly created and still soft earth, right in the center of the snow peaks of the Sierra Nevada, saying "This is the central post", and then picking from the top of the spindle a length of yarn, she drew with it a circle around the spindle-whorl and said "This shall be the land of my children!"'

A Kogi spindle, then, is a model of the cosmos: the flat disk of the spindle-whorl is our Earth and on top of it rests the high, cone-shaped body of cotton yarn wound tightly around the world axis.

On occasion this whorl is decorated with four little dots engraved on opposite sides of the circular object. They represent the World-Quarters, or else [the World-Quarters are designated by] two incised intersecting lines in the form of a cross. The white yarn is 'the thought of the sun'; it represents life light—a male seminal concept of fertility and growth.

The white cone is seen as divided horizontally into four ascending levels that represent the Upper World. At the summit is the sun. Underneath the world disk is another, inverted, cone of yarn, and the Kogi talk of an invisible cone of black thread, also divided into four levels that represent the Lower World: 'The sun, by spiralling around the world, spins the Thread of Life and twists it around the cosmic axis; during the day a left-spun white thread and during the night a right-spun black one'.[131]

Weaving plays an important role in the social and spiritual lives of many Central American peoples, including the Maya in the highlands of Chiapas. The Maya weaving tradition is a very old one. Most ancient Maya art forms did not survive the collapse of Classic Maya society in the tenth century, let alone the Spanish Conquest in the sixteenth.

Mayan woven textiles were exceptionally beautiful and elaborate. As Walter F. Morris notes, the huipal, a rectangular blouse that is the ceremonial costume of the ancient and modern Maya, 'is woven with designs that symbolize their vision of the cosmos and the beings that bring rain and life to the world'. The weaver is enveloped 'in a diagram of the world about to flower'. Morris provides a beautiful description of the weaver in her ceremonial finery with all its cosmic imagery:

> Radiating from her head are diamonds that depict the sun's movement through the sky and underworld. Along the edges of the brocade the musicians of rain, the toads, dance with the Earthlord who creates the clouds and reigns over the flowering plants that appear in growing rows of designs that cover the sleeves. A weaver may include depictions of the ancestors, the

patron saints, the people of the first world who became monkeys, the monsters defeated by the Earthlord, and the animal spirits with their jaguar lord. She interweaves these designs and the power they symbolize into a harmonious vision of the world renewed in flower.[132]

Very much the same is true of weaving in the islands of the Sumba group that are at present part of Indonesia. The anthropologist Danielle Geirnaert-Martin has made an exhaustive study of the cosmic symbolism in the textiles produced on one of the islands called Laboya. For the Laboyans, the weaving of cloth is seen as essential for maintaining the stability and order of a woman's life, that of her family, and that of the natural world and the all-encompassing cosmos.

There is an analogy between spinning and birth: the ball of thread symbolises a foetus or a newborn baby. It can be made into warp and weft and then woven: for the process of weaving cloth and rearing children is one of equivalence. The positions women take while weaving is said to be good for coitus and bearing children. It is most important to ensure that the warp threads are evenly spaced before beginning. If they are not even, the woman may have miscarriages, the waters of the land may flow out, and the earth may dry up. If the threads cross, it is likened to incest. In a newborn, the elements of the soul need to be welded together using a loom swift, in the same way that it is used to wind yarn from a skein into a ball, otherwise the mawo (breath, life force) may become restless and leave the body. The swift also represents the link between Heaven and Earth, between the living and the dead.[133]

As with the Kogi, the rationale for such beliefs is found in their creation myths. In one myth the Laboya ancestor, Ubu Raba, took the form of a python and wove the land into existence. He wove the fountains, beating the weft into the warp with the sword until all springs were well enclosed by the woven land. This is how the earth grew moist and young again. Human beings are said to have originated by being plaited or spun, and the moon is responsible for welding together their physical and spiritual properties. The Laboyans divide the sea and the land into seven layers each. Weaving restores the seasons and the fertility of land, animals and humans. Ubu Raba established order within society by passing the shuttle in the correct direction, and he set out conditions for obeying the rules as he wove the cloth of the world.

On the Laboya loom, the warp is continuous: the warp beam is attached to the horizontal tie-beam above the floor, and the bottom beam is tied to the weaver's back at the waist. According to Geirnaert-Martin, a Laboyan who weaves in the 'proper' order ensures 'a correct relationship with above and below, with the Earth and the Sky':

> It is only then that the cycle of the celestial bodies and the alternations of the seasons is resumed. Uba Raba created life using a continuous warp and this is a metaphor for the human life cycle. Cotton is likened to woman's breath or life force. As the original ancestor, Ubu Raba wove cosmic order, the natural environment, and the possibility for regeneration.[134]

Laboyans weaving their traditional textiles are also weaving the cosmos—thereby maintaining its critical order. If they fail to perform the sacred rites involved, they are violating the sacred laws that govern their society, the natural world, and indeed the all-encompassing cosmos. To weave in any other way would, in many traditional societies, be regarded as taboo. In the words of Roger Caillois, 'an act is taboo because it disrupts the universal order, which is at once that of nature and society'. If that order is disrupted, 'the Earth might no longer yield a harvest, the cattle might be struck with infertility, the stars might no longer follow their appointed course, death and disease could stalk the land'. To violate a taboo is to be guilty of cosmic sin.[135]

This may appear to be true. The recent storms and floods in Orissa and Vietnam, and the increased incidence of devastating droughts throughout the world, are the result of cutting down forests and of transforming the chemical composition of the atmosphere and hence disrupting the order of the ecosphere. Whether we like it or not, the religio-culture of tribal peoples tells them truths about their relationships with the cosmos. These truths are imparted in a way understood and believed in not just intellectually, but through heart and soul, in the way that is most likely to be acted upon.

Although these ideas figured prominently in the theologies of our early mainstream religions, we now have lost sight of them. Cosmic consciousness must be resuscitated, for only in this way can religion and the arts inspire people to unite against the forces of chaos that today are threatening our very survival.

*

In the discussion which followed Edward Goldsmith's talk, one exchange seemed especially pertinent to the wider issues addressed:

Q. I am concerned that you have come across as being against progress. Surely the question is one of scale? Nowadays aspects of life in pre-industrial times are magnified a hundred or thousand-fold and so have become totally out of balance—isn't that the issue?

A. It is, but it is more than that. You see I *am* against progress. I have to admit it, and I am sure that is not the way to become popular . . .. The obvious manifestation of progress is economic development. Let us examine one aspect of what we call economic development. If you look at a traditional society, the family and the community fulfil almost all the functions that satisfy the requirement of basic needs. In our society corporations and the state fulfil these needs. In the traditional society the community produced food and cooked it, looked after the old and young, educated into the culture of the society, organized and performed the religious rituals that were required and governed themselves—until quite recently in Switzerland, for instance, political power theoretically resided in the village or commune, the residual power only going to the canton. They did not produce food or artefacts to satisfy economic requirements.

All these things were fulfilled without money. Carl Polanyi, the great economic historian, has pointed out that the formal economy is something new. In these traditional societies economic relations were totally embedded in social relations. If you look at the definitions of poverty in the First World, all of them involve money. In our society everything is going to become commercialized, we are removing all social relationships, nothing will be done for free. This is a very unstable and hopeless situation. Economic development means dis-embedding functions that have always been fulfilled at a family and community level from their social context. It means monetizing and commoditizing these functions so that we become dependent on the economy. The economy is collapsing now, and without the money to fulfil these functions it will vanish. You can see today that we are dismantling the welfare state slowly but systematically. Jobs are becoming precarious, with short-term contracts etc., and we will have to return to what was the original source of security, which was provided by membership of a family and a community. Traditional peoples *understand* this. The closest thing to the meaning of the word 'poor' in West Africa is the word 'orphan': someone who is deprived of social relations.

DAVID CADMAN
With Our Thoughts We Make the World

🍂

EDITORS' NOTE: *David Cadman gave this as the final lecture of the series, speaking in his capacity as Chairman of the Prince's Foundation.*

Nearly thirty years ago, in his now classic treatise *Small is Beautiful: A Study of Economics as if People Mattered*, Fritz Schumacher made the following observations in a chapter entitled 'The Proper Use of Land':

> Study how a society uses its land, and you can come to pretty reliable conclusions as to what its future will be . . ..

> . . . 'The Proper Use of Land' poses, not a technical nor an economic, but primarily a metaphysical problem.

> In the simple question of how we treat the land . . . our entire way of life is involved, and before our policies with regard to the land will really be changed, there will have to be a great deal of philosophical, not to say religious, change.[136]

And so, in such a quest, in the journey to unravel this dilemma of values, I take as my guide the opening lines of the Buddhist *Dhammapada*, which are as follows:

> We are what we think
> All that we are arises with our thoughts
> With our thoughts we make the world.[137]

And again, in another translation:

> What we are today comes from our thoughts of yesterday, and our present thoughts build our life of tomorrow: our life is the creation of our mind.[138]

In this way, it seems to me that the language that we use to describe our relationship with the land is made concrete in all that we build, in the way that we farm, and in the ways in which we care for the Earth. What we do is a function of what we think, and thought, captured in the beguiling snare of language, defines and can constrain, and even distort, our reality.

For language, laden as it is with explicit and implicit values, does more than describe, it *governs*. Thus it is that it gives meaning and worth and, thereby, defines and circumscribes that which we take to be real. In this sense, 'reality' is not absolute but dependent, dependent upon the language that we use to describe it. Furthermore, it is not simply dependent upon our own particular and chosen language but, most importantly, dependent upon a language that others have framed and that, knowingly or, more probably, unknowingly, we have accepted.

Thus an engineer, a banker and a poet will survey the same landscape and see it quite differently from each other. To one, it is configured according to its geological form; to another it is property and collateral; and to the third it calls forth memory and imagination. With our thoughts we make the world.

If, then, we find that much of what we build is mean, ugly and out of place; if we find that our food and water are degraded and unclean; if we come to burn our cattle on funeral pyres and pillage the seas until they are exhausted; if we fear for the Earth and doubt our capacity to live in harmony with the rest of nature; we cannot find our way out of the dilemma other than by challenging the very language and values that have brought us to where we are.

Our present language, the language that has come to shape our world, has, perhaps, two main characteristics. Firstly, it is *reductionistic*, which is to say that it is a language of parts and not of wholes. It is a language that encourages us to see ourselves as 'separate from' rather than as 'a part of'. And, secondly, it is overwhelmingly a language of *economics, accounting and finance*, which is to say that it is a language of prices. It supposes that those things that are real, and of real value, are only those that can be measured by price. By contrast it implies that those things that are not readily priced are in some sense unreal and of no real value. Whatever the matter, at every place, we are urged to be 'hard-nosed' and to focus upon 'the bottom line', as though our whole lives could be measured in the columns of profit and loss and all that we value captured in a balance sheet.

Most especially, this language is based upon the proposition that what is expressed as 'the *real* world' is that which is tangible, concrete and fixed. It is made up only of those things that are capable of being measured, possessed and consumed. And the real dilemma is that this language, useful and productive in its own way but nevertheless limited in its relevance and usefulness, has come to be taken as having universal application. It is applied not simply to some kinds of science and to some kinds of market transaction but to all that we do. It has come to define and limit our experience. It is everywhere and governs all.

Thus, to paraphrase the *Dhammapada*: 'With our thoughts [of separateness] we make the world.'

And: '. . . our life [has become] the creation of our [accounting] mind.'

It seems to me, then, that if we are to check, or even reverse, the present degradation of the web of life we must challenge the universal application of this language: put it back in its place as a useful but partial language, and seek another way of being. And might it be that this 'other way' can be 'a sacred trust'? I believe this to be so, and I am encouraged in this conclusion by the words of Seyyed Hossein Nasr:

> the environmental crisis . . . requires a very radical transformation in our consciousness, and this means not discovering a completely new state of consciousness, but returning to the state of consciousness that traditional humanity always had. It means to rediscover the traditional way of looking at the world of nature as sacred presence.[139]

It should be noted that Nasr refers to 'the traditional way of looking at the world'. And since it is HRH's insistence upon the importance of 'tradition' that is most often ridiculed and misrepresented, it is perhaps worth emphasizing that in the context of 'the sacred' it has particular meaning. Here, and I believe in much that HRH has to say, the word 'tradition' is not *primarily* a matter of an aesthetic but rather an 'outlook' and a 'way of being', a way of being that is based upon reverence and upon compassion, and an outlook that sees in the manifest world of nature a glimpse of the mysterious and awesome order and harmony of the Divine. In this sense, the word 'tradition' is not determined by time, but by a set of values which testify to the existence of timeless principles—a living and not a dead tradition. This tradition is not 'of man' but 'of God'. And, ironically, this 'tradition' is, today, a radical, even dissident, proposition and one that is most upsetting to those who would have us accept the narrow and dismal confines of their exclusively secular world!

The 'tradition' to which I refer is not, of course, that which is represented by the sentimentality of nostalgia or pastiche, but bears the much more profound sense that Marco Pallis refers to in his book *The Way and the Mountain*. In a discussion of the relationship of the Active Life and the Contemplative Life, he contrasts the 'traditional outlook' with the 'anti-traditional outlook' in this way:

> ... whereas the traditional outlook fosters a habit of always looking to the cause rather than to the effect in all orders ... the anti-traditional [and we might say modernist] attitude encourages precisely the contrary tendency, namely the paying of more attention to applications than to principles, to effects than to causes, to symptoms rather than to the disease .... *This mental habit, which is all the more dangerous in that it is largely unconscious, lies at the root of most of our troubles* ... (my emphasis).[140]

It is clear, then, if you follow this argument, that we have to change our ways; and to do this we will surely have to learn to see more clearly. For since it is the case that we make our world with our thoughts, since there is an unbroken link between what we do and what we believe to be true, it seems to me that we have to accept that our dilemma is rooted in ignorance. In an age of hubris this is not at all easy. Especially it is not easy for those who have been encouraged to believe that they know what is true and, indeed, what is best for us. In this, I cannot help but quote the rather chilling words of a scientist responding last year to the thoughts of HRH in the Reith Lecture: 'People say we are playing God. But if scientists don't play God, who will?'[141]

Who indeed!

Interestingly enough, by contrast, the object of the criticism, the Prince himself, had spoken of our need for humility: 'Our most eminent scientists accept that there is still a vast amount that we don't know about our world and the life forms that inhabit it . . .'. Indeed,

> Faced with such unknowns, it is hard not to feel a sense of humility, wonder and awe about our place in the natural order. And to feel this at all stems from that inner, heartfelt reason which, sometimes despite ourselves, is telling us that we are intimately bound up in the mysteries of life and that we don't have all the answers.[142]

The teachings on humility, on reverence, and on the virtue of what in Zen is called 'beginner's mind', are a part of all the great spiritual traditions. It is also a part of these teachings that dispelling ignorance requires a mind that is clear, attentive and still; that contemplation is the root of true action; that meditation and mindfulness are the path to wisdom. And linked to this, it is clearly taught that wisdom cannot be attained other than by practice. The parables of Jesus were always given in terms of everyday experience; and in Buddhism, too, it seems to me, we are offered not only a way of knowing but also a way of being, a way of engaging with the world:

> Subhuti asked [the Buddha]: 'Is it possible to find perfect wisdom through reflection or listening to statements or through signs or attributes, so that one can say "This is it" or "Here it is"?'

The Buddha answered: 'No, Subhuti. Perfect wisdom cannot be learned or distinguished or thought about or found through the senses. This is because nothing in this world can be finally explained, it can only be experienced, and thus all things are just as they are. Perfect wisdom can never be experienced apart from all things. To see the Suchness of things, which is their empty calm being, is to see them just as they are. It is in this way that perfect wisdom and the material world are not two, they are not divided. As a result of Suchness, of calm and empty being, perfect wisdom cannot be known about intellectually. Nor can the things of the world, for they are understood only through names and ideas. Where there is no learning or finding out, no concepts or conventional words, it is in that place one can say there is perfect wisdom.'[143]

In this way, then, and in this search for a new language and a new way of being, it would seem that to talk of a sacred trust is not to talk of something that is esoteric and apart. Rather it is to engage most deeply, most immediately and most entirely with the world. This focus upon heartfelt action, upon a *practical* response that meets the deepest needs of all people and of all beings, lies very much at the heart of the Prince's Foundation and indeed, I believe, of all that HRH proclaims.

In exploring the values that lie beneath our present environmental crisis, I have so far spoken of the causal relationship between our thoughts and our deeds, and therefore of the importance of our way of being, and of the need for mindfulness and attentiveness bedded within practical action. In particular, however, I have also spoken of the need to challenge the language that at present shapes the way in which we build and make communities, for it is my belief that the experience that this language defines is narrow, particular and flawed.

Firstly, in proposing *a reality of parts* it fails to provide us with a sense of the whole. As the Buddhist scholar Stephen Batchelor puts it: 'This habit of isolating things leads us to inhabit a world in which the gaps between them become absolute.'[144] Thus it is that we are bereft of meaning.

Secondly, in proposing *a reality of transactions*, this language lacks compassion. In focusing upon rights rather than responsibilities, it distorts and even denies relationships. In failing to take account of its own 'external' costs, it is often blind to the degradation and misuse of nature.

Given its narrow frame of reference and evident capacity for humbug, let alone failure, it is bizarre that this language, this reality, should hold such sway. But it does; and it has coarsened our touch, dulled our eye and, most importantly, taken our tongue. In many ways, it would seem, we are, quite literally, lost for words.

Finally, though, I would like to add something more on 'relatedness'. A common theme in all the talks in this series *Ecology: A Sacred Trust* is that we must come to see the world as related, connected and whole; that all that is is part of an intricate web of causation and dependency; and, indeed, that we should see ourselves as 'a part of' and not 'apart from'. I have already spoken of the relationship between our thoughts and our actions; and perhaps the most compelling idea of our time is the rediscovery of a *reality of relatedness*. In physics, in biology, in economics, in medicine, in the arising of the entire debate about sustainability, the limitations of reductionism and of the science of parts are being exposed. So much has been written on this new reality that for the purpose of this essay I take it as read. But the challenge now is to know how to proceed in a more integrated way, recognizing and responding to the reality of contingency and reciprocity.

I should add, of course, that such a notion of relatedness is not in the least bit new. It was, after all, the Greek philosopher Heraclitus who saw reality as an ever-changing river where everything flows.[145] It is there in the words of Meister Eckhart and the Christian mystics. And, most especially, in the form of *paṭiccasamuppāda* or 'dependent origination', it lies at the very heart of the teachings of the Buddha.

For it is taught: 'One who sees dependent origination sees the Dhamma; one who sees the Dhamma sees dependent origination.'[146]

Or, again: 'That is when this is; that arises with the arising of this. That is not when this is not; that ceases with the cessation of this.'[147]

Such a notion is material for another lecture. Nevertheless, so much of our education has been counter to this view that we need to learn not just new skills but a new way of seeing the world.

At one level, for the Prince's Foundation, this means developing new approaches to urban regeneration—meeting the social, economic, environmental, spiritual and cultural needs of people in a coherent way—and developing the connection between our programmes of education and our portfolio of projects. It also means working to establish links beyond our boundaries by the creation of a new Network for Urban Affairs, which will be a 'one-stop shop' providing information about, and giving on-line access to, practitioners, products and projects that share the over-riding themes of integration, 'organic' urbanism and the celebration of traditional building crafts.

But beyond this, I am left with a feeling both of profound excitement and unease. I feel like someone standing at the very beginning of a journey, and hearing what Joseph Campbell refers to as 'the call to adventure':[148]

> The first stage of the mythological journey [the 'call to adventure'] signifies that destiny has summoned the hero and transferred his spiritual center of gravity from within the pale of his society to a zone unknown. This fateful region of both treasure and danger may be variously represented: as a distant land, a forest, a kingdom underground, beneath the waves, or above the sky, a secret island, lofty mountain top, or profound dream state . . .. The adventure may begin as a mere blunder . . . or still again, one may be only casually strolling, when some passing phenomenon catches the wandering eye and lures one away from the frequented paths of men.[149]

In a world so taken up with rationality and intent, with business plans and five-year strategies, is it not especially thrilling to be told that 'the adventure' is always there and always about to begin; that we might stumble upon it when we least expect to or 'when some passing phenomenon catches [our] wandering eye and lures [us] away from the frequented paths of men'?

And so it is that I feel that this exploration of the reality of related-ness will take us to places of which we have little or no knowledge, to 'fateful region(s) of both treasure and danger', as Campbell puts it. But, at the same time, I have a sense that these realms will seem familiar when we arrive. Many of our present opinions will be chal-lenged but we may well recognize the ground upon which we come to stand: for our past, our present and our future are not discon-nected events but chapters in a single and unfinished story.

I hope that this series *Ecology: A Sacred Trust* will have encouraged us to accept the 'call to adventure', to have the courage and the determination to challenge convention, in the words of HRH, to 'rediscover a reverence for the natural world'[150] and to be attentive to where this takes us—in all that we do.

And so we come to the end of the beginning. In his great teaching on relatedness, the Buddhist sage Nāgārjuna said:

> I bow to buddhas
> Who teach contingency
> . . .
> And ease fixations.[151]

If we are to tackle our present ecological dilemma as if the task were a sacred trust, will we have the courage to ease fixations, like the Grail seekers taking the path of uncertainty and entering the darkest part of the forest? Or will we cling to the comfort of what we have come to accept and keep our fingers crossed? Who can say? But, as the *Dhammapada* tells us, of one thing we can be sure:

> What we are today comes from our thoughts of yesterday, and our present thoughts build our life of tomorrow: our life is the creation of our mind.

KATHLEEN RAINE

Millennial Hymn to the Lord Shiva[152]

❧

1

Earth no longer
Hymns the Creator,
The seven days of wonder,
The Garden is over –
All the stories are told,
The seven seals broken
All that begins
Must have its ending,
Our striving, desiring,
Our living and dying,
For Time, the bringer
Of abundant days
Is Time the destroyer –
    In the Iron Age
    The Kali Yuga
    To whom can we pray
    At the end of an era
    But the Lord Shiva,
    The Liberator, the purifier?

2

Our forests are felled,
Our mountains eroded,
The wild places
Where the beautiful animals
Found food and sanctuary

We have desolated,
A third of our seas,
A third of our rivers
We have polluted
And the sea-creatures dying.
Our civilisation's
Blind progress
In wrong courses
Through wrong choices
Has brought us to nightmare
Where what seems,
Is, to the dreamer,
The collective mind
Of the twentieth century –
This world of wonders
Not divine creation
But a big bang
Of blind chance,
Purposeless accident,
Mother Earth's children,
Their living and loving,
Their delight in being
Not joy but chemistry,
Stimulus, reflex,
Valueless, meaningless,
While to our machines
We impute intelligence,
In computers and robots
We store information
And call it knowledge,
We seek guidance
By dialling numbers,
Pressing buttons,
Throwing switches,
In place of family

Our companions are shadows,
Cast on a screen,
Bodiless voices, fleshless faces,
Where was the Garden
A Disneyland
Of virtual reality,
In place of angels
The human imagination
Is peopled with footballers,
Filmstars, media-men,
Experts, know-all
Television personalities,
Animated puppets
With cartoon faces –
 To whom can we pray
 For release from illusion,
 From the world-cave,
 But Time the destroyer,
 The liberator, the purifier?

3

The curse of Midas
Has changed at a touch,
A golden handshake
Earthly paradise
To lifeless matter,
Where once was seed-time,
Summer and Winter,
Food-chain, factory farming,
Monocrops for supermarkets,
Pesticides, weed-killers,
Birdless springs,
Endangered species,
Battery-hens, hormone injections,

Artificial insemination,
Implants, transplants, sterilization,
Surrogate births, contraception,
Cloning, genetic engineering, abortion,
And our days shall be short
In the land we have sown
With the Dragon's teeth
Where our armies arise
Fully armed on our killing-fields
With land-mines and missiles,
Tanks and artillery,
Gasmasks and body-bags,
Our aircraft rain down
Fire and destruction,
Our spacecraft broadcast
Lies and corruption,
Our elected parliaments
Parrot their rhetoric
Of peace and democracy
While the truth we deny
Returns in our dreams
Of Armageddon,
The death-wish, the arms-trade,
Hatred and slaughter
Profitable employment
Of our thriving cities,
The arms-race
To the end of the world
Of our postmodern, post-Christian,
Post-human nations,
Progress to the nihil
Of our spent civilisation.
But cause and effect,
Just and inexorable
Law of the universe

No fix of science,
Nor amenable god
Can save from ourselves
The selves we have become,
We are all in it,
No one is blameless –
   At the end of history
   To whom can we pray
   But to the destroyer,
   The liberator, the purifier?

4

In the beginning
The stars sang together
The cosmic harmony,
But Time, imperceptible
Taker-away
Of all that has been,
All that will be,
Our heart-beat your drum,
Our dance of life
Your dance of death
In the crematorium,
Our high-rise dreams,
Valhalla, Utopia,
Xanadu, Shangri-la, world revolution
Time has taken, and soon will be gone
Cambridge, Princeton and M.I.T.,
Nalanda, Athens and Alexandria
All for the holocaust
Of civilisation –
   To whom shall we pray
   When our vision has faded
   But the world-destroyer,
   The liberator, the purifier?

5

But great is the realm
Of the world-creator,
The world-sustainer
From whom we come,
In whom we move
And have our being,
About us, within us
The wonders of wisdom,
The trees and the fountains,
The stars and the mountains,
All the children of joy,
The loved and the known,
The unknowable mystery
To whom we return
Through the world-destroyer –
    Holy, holy
    At the end of the world
    The purging fire
    Of the purifier, the liberator!

Notes

❧

1. This lecture was organized by the Stephen Lawrence Charitable Trust and was given to an invited audience at the Prince's Foundation on 7 September 2000.

2. HRH The Prince of Wales, 'A Reflection on the Reith Lectures for the Year 2000', *Temenos Academy Review* 4 (Spring 2001), pp. 13–18: p. 15. The quotation is taken from Philip Sherrard, *Christianity: Lineaments of a Sacred Tradition* (Brookline, Massachusetts: Holy Cross Orthodox Press and Edinburgh: T & T Clarke, 1998), p. 243.

3. This essay was first published by the Temenos Academy as Temenos Academy Papers No. 4 in 1995, and republished in 1998 as 'The Desecration of the Cosmos' in *Christianity: Lineaments of a Sacred Tradition*, pp. 200–31. For its reprinting in the present volume, a few references have been supplied by the editors. The Temenos Academy wishes to acknowledge its gratitude to the Holy Cross Orthodox Press for permission to re-publish this paper, and also to the Literary Executors of the Estate of Philip Sherrard, Denise and Liadain Sherrard. The essay's title is the last line of William Blake's 'The Marriage of Heaven and Hell'.

4. Fragment 107.

5. Thus cited by Aristotle, *Metaphysics* 1053a.

6. *Spiritual Espousals* 3.7.

7. 2 Corinthians 3:6.

8. *Confessions* 9.10.

9. Matthew 25:40.

10. Christopher Marlowe, *Doctor Faustus*, Prologue, line 20.

11. *Taittirīya Upaniṣad* 2.2; trans. Robert Ernest Hume, *The Thirteen Principal Upanishads* (2nd English ed.: Oxford University Press, 1931), p. 284.

12. Ibid. 3.2; trans. Hume, p. 290.

13. Ibid. 2.2; trans. Hume, p. 284.

14. *Mahā Aśvamedhika* 92.

15. *Taittirīya Brāhmaṇa* 2.8.8.1.

16. Vandana Shiva, *Tomorrow's Biodiversity* (London and New York: Thames and Hudson, 2000).

17. 'Poverty and Globalisation', BBC Reith Lecture, 10 May 2000.

18. For a scholarly study of Jain philosophy the reader can consult the new translation of the *Tattvārtha* of Umāsvāti, entitled *That Which Is: A Classic Jain Manual for Understanding the Nature of Reality*, trans. Nathmal Tatia, with an extensive introduction by Padmanabh S. Jaini (Walnut Creek CA: AltaMira Press, 1994).

19. James Lovelock, *Homage to Gaia: The Life of an Independant Scientist* (Oxford: Oxford University Press, 2000), pp. 412–19.

20. N. Fettling, 'Mildura Palimpsest: A Confluence of Science and Art', *Ecopolitics* 1 (2000), pp. 18–23.

21. For further discussion, see R. V. Solé and B. Goodwin, *Signs of Life: How Complexity Pervades Biology* (New York: Basic Books, 2000).

22. See F. Wemelsfelder, E. A. Hunter, M. T. Mendl and A. B. Lawrence, 'The Spontaneous Qualitative Assessment of Behavioural Expressions in Pigs: First Exploration of a Novel Methodology for Integrative Animal Welfare Measurement', *Applied Animal Behaviour Science* 67 (2000), pp. 193–215, for studies on the quality of experience of pigs, observed directly from their behaviour.

23. J. Naydler, trans., *Goethe on Science: An Anthology of Goethe's Scientific Writings* (Edinburgh: Floris Books, 1996), p. 116.

24. More information on the relationship between Bahá'í principles and the environment can be found in the Bahá'í International Community's statement presented to the Preparatory Committee of the United Nations Conference on Environment and Development (UNCED). See Bahá'í International Community, 'Earth Charter', 5 April 1991, which can be found at http://www.bic-un.bahai.org/92-0606.htm (accessed on 16 June 1999).

25. 'Abdu'l-Bahá as quoted in H. M. Balyuzi, *'Abdu'l-Bahá: The Centre of the Covenant of Bahá'u'lláh* (Oxford: George Ronald, 1987), p. 208.

26. 'Abdu'l-Bahá, *The Will and Testament of 'Abdu'l-Bahá* (Wilmette, Ill.: Bahá'í Publishing Trust, 1944), pp. 13–14.

27. Shoghi Effendi, *The World Order of Bahá'u'lláh* (Wilmette, Ill.: Bahá'í Publishing Trust, 1982), p. 40.

28. Winchester Celebration, 'Creation Harvest Liturgy: Order of Ceremony' (Winchester Cathedral: World Wildlife Fund, 4 October 1987), pp. 8–9.

29. The concept of a 'world federal system' is derived from that which is stated by Shoghi Effendi, *The World Order of Bahá'u'lláh*, pp. 39–41.

30. Genesis 1:28.

31. 'Abdu'l-Bahá, *Selections from the Writings of 'Abdu'l-Bahá*, trans. Committee at the Bahá'í World Centre and Marzieh Gail (Haifa: Bahá'í World Centre, 1978), pp. 193–4.

32. Bṛhadāraṇyaka Upaniṣad 2.4.5, trans. Juan Mascaró, *The Upanishads* (Harmondsworth: Penguin, 1965), p. 131.

33. Bahá'u'lláh, *Tablets of Bahá'u'lláh revealed after the Kitáb-i-Aqdas*, trans. Habib Taherzadeh with the assistance of a committee at the Bahá'í World Centre (Wilmette, Ill.: Bahá'í Publishing Trust, 1988), p. 142.

34. Bahá'u'lláh, *Gleanings from the Writings of Bahá'u'lláh*, trans. Shoghi Effendi (Wilmette, Ill: Bahá'í Publishing Trust, 1952), p. 262.

35. Ibid., p. 178.

36. Bahá'u'lláh, *Epistle to the Son of the Wolf*, trans. Shoghi Effendi, rev. ed. (Wilmette, Ill.: Bahá'í Publishing Trust, 1953), p. 44.

37. Bahá'u'lláh, *The Hidden Words*, trans. Shoghi Effendi (Manchester: Bahá'í Publishing Trust, 1949), §48 from the Persian, p. 40.

38. Ibid., §29 from the Persian, p. 33.

39. Clement of Alexandria, *Protrepticus*, chapter 2; edited in J.-P. Migne, *Patrologia Graeca* viii. 77–8.

40. Qur'án 7.56.

41. From a Pazand general confession, as cited in Mary Boyce, ed. and trans., *Textual Sources for the Study of Zoroastrianism* (Totowa, New Jersey: Barnes & Noble, 1984), p. 60.

42. From a card series produced by the Bahá'í International Community, 1986.

43. 'Abdu'l-Bahá, *Selections*, pp. 158–9.

44. Bahá'u'lláh, *Gleanings*, p. 97.

45. Ibid., p. 84.

46. Bahá'u'lláh, *Kitáb-i-Íqán: The Book of Certitude*, trans. Shoghi Effendi, (Wilmette, Ill.: Bahá'í Publishing Trust, 1950), p. 191.

47. Bahá'u'lláh, *Prayers and Meditations by Bahá'u'lláh*, trans. Shoghi Effendi (Wilmette, Ill.: Bahá'í Publishing Trust, 1962), p. 272.

48. Bahá'u'lláh, *Gleanings*, p. 43.

49. Ibid., p. 30.

50. Qur'án 24.45.

51. Ibid., 24.35.

52. Part I, line 2039.

53. 'Abdu'l-Bahá, *Selections*, p. 135.

54. The Báb, *Selections from the Writings of The Báb*, trans. Habib Taherzadeh with the assistance of a committee at the Bahá'í World

Centre (Haifa: Bahá'í World Centre, 1976), p. 95.

55. Universal House of Justice, letter dated 31 August 1987 addressed to the Bahá'ís of the world, quoted in 'Mountain of the Lord: The Terraces and the Arc' (Bahá'í Publishing Trust, n.d.).

56. Shoghi Effendi, letter dated Naw-Rúz 108 (21 March 1952) addressed to the Bahá'ís in the East, quoted in ibid.

57. Ibid.

58. Bahá'u'lláh, Gleanings, p. 160.

59. Leo R. Zrudlo, 'The Missing Dimension in the Built Environment', Journal of Bahá'í Studies 3.1 (1990), pp. 49–57.

60. 'Abdu'l-Bahá, The Secret of Divine Civilization, trans. Marzieh Gail in consultation with Ali-Kuli Khan (Wilmette, Ill.: Bahá'í Publishing Trust, 1990), p. 35.

61. Ibid., p. 39.

62. Ibid., p. 39.

63. Ibid., pp. 24–5.

64. Ibid., p. 24.

65. Juvenal, Satire 3.190–211.

66. 'Abdu'l-Bahá, The Secret of Divine Civilization, p. 4.

67. Ibid., pp. 3–4.

68. 'Abdu'l-Bahá, Selections, pp. 158–60.

69. Elsewhere, 'Abdu'l-Bahá explains that such animals 'in reality as regards themselves . . . are good'. Their only 'evil' is in the 'antagonism' between certain of their elements and our elements which 'do not agree'. See Some Answered Questions, collected and translated from the Persian by Laura Clifford Barney (Wilmette, Ill.: Bahá'í Publishing Trust, 1981), pp. 263–4.

70. 'Abdu'l-Bahá, Selections, p. 153.

71. 'Abdu'l-Bahá, The Will and Testament of 'Abdu'l-Bahá, pp. 13–14.

72. 'Abdu'l-Bahá, The Secret of Divine Civilization, p. 64.

73. 'Abdu'l-Bahá as quoted in Shoghi Effendi, The World Order of Bahá'u'lláh, p. 39.

74. Shoghi Effendi, ibid., p. 41

75. Bahá'u'lláh, Gleanings, p. 250.

76. Shoghi Effendi, The World Order of Bahá'u'lláh, p. 41.

77. See Bahá'í International Community, 'International Legislation for Environment and Development', 5 April 1991, which can be found at http://www.bic-un.bahai.org/92-0222.htm (accessed on 16 June 1999).

78. Asclepius 8; translation adapted from Walter Scott, Hermetica: The Ancient Greek and Latin Writings which contain Religious or Philosophical Teachings ascribed to Hermes Trismegistus, 4 vols (Oxford: Clarendon Press, 1924–36; reissued Boston: Shambhala, 1983–5), i.300–1. A condensed version of Scott's edition has recently been published with a foreword by A. G. Gilbert (Bath: Solos Press, 1992): the passage quoted above appears there on p. 121.

79. Aristotle, Nicomachean Ethics 1115a–1117b.

80. Matthew 4:1–11; cf. Mark 1:12–13, Luke 4:1–13.

81. Matthew 4:4.

82. Cited in Richard Tarnass, The Passion of the Western Mind (New York: Ballantine, 1991), p.308 and p.488 n. 12.

83. For a recent discussion of the 'computational theory of mind', see Steven Pinker, How the Mind Works (New York: Norton, 1998).

84. Brenda Walpole, The Human Machine (Hove: Wayland, 1990), p. 5.

85. For an excellent account of Bacon's view of the purpose of scientific research, see Theodore Roszak, Where the Wasteland Ends: Politics and Transcendence in Post-industrial Society (London: Faber and Faber, 1973), Chapter 5.

86. Despite the UK Government's rejection of the proposal on 16 July 1999, the pressure to install a system of 'presumed consent' to replace the current 'opt in' system of donor cards is bound to intensify as the

demand for organs increases. See *The Independent*, 17 July 1999, p. 5.

87. *The Guardian* (Society), 5 March 1997, p. 4.

88. *The Times*, 31 January 2001, p. 8.

89. *The Times*, 16 July 1999.

90. Paul Pearsall, *The Heart's Code* (London: Thorsons, 1998), Chapter 4.

91. Kevin Warwick, *The March of the Machines* (London: Century, 1997), p. 146.

92. *The Times*, 15 March 2000, p. 13.

93. The guidelines were in a report drawn up by the Government-sponsored infection surveillance steering group of the UK Xeno-transplantation Interim Regulatory Authority (UKXIRA), and appear on the front page of *The Daily Telegraph*, 25 October 1999.

94. Naydler, *Goethe on Science*, p. 112.

95. *The Times*, 12 January 2001, p. 13.

96. *The Times*, 24 July 2000, p. 8.

97. Ibid.

98. Patrick Dixon, *The Genetic Revolution* (Eastbourne: Kingsway Publications, 1995), pp. 89ff. For the new salmon that grows six times as fast as normal, see *The Times*, 12 April 2000.

99. Matthew Wenban-Smith, 'Something Nasty in the Woodshed?', *Living Earth* (*Magazine of the Soil Association*) (April/June 1999), pp. 1of.

100. Dixon, op. cit. p. 76.

101. Jeffrey Burton Russell, *The Devil* (New York: Cornell University Press, 1977), p. 34.

102. *The Times*, 20 March 2000, p. 3.

103. See the excellent study by Robert Romanyshyn, *Technology as Symptom and Dream* (London: Routledge, 1989).

104. *The Times*, 2 October 1999, p. 15.

105. This statistic is from the 2001 edition of *Social Trends*, published by the Office for National Statistics, and reported in *The Independent*, 25 January 2001, p. 12.

106. There are several versions of this story, one of which appears in the Qur'ān (7.11-18). Among Jewish and Christian sources, one of the oldest is that in *Vita Adae et Evae* 13-16: see James H. Charlesworth, ed., *The Old Testament Pseudepigrapha*, 2 vols (Garden City, New York: Doubleday, 1983-5), ii.262.

107. *The Times*, 25 November 2000, p. 13.

108. A. N. Whitehead, *Religion in the Making* (Cambridge: Cambridge University Press, 1926), p. 5.

109. *The Independent*, 12 July 2001, pp. 6-7.

110. *Saṃyutta Nikāya* 47.20, quoted in Thich Nhat Hanh, *The Miracle of Mindfulness* (London: Ryder, 1991), p. 63.

111. This essay was first published by the Temenos Academy as Temenos Academy Papers No. 12 in 1999.

112. René Guénon, *The Crisis of the Modern World*, trans. Marco Pallis and Richard Nicholson (London: Luzac, 1962); originally published in 1927 as *La crise du monde moderne*.

113. William C. Chittick, *The Self-Disclosure of God: Principles of Ibn al-'Arabī's Cosmology* (Albany: State University of New York, 1998).

114. Qur'ān 17.44 (Pickthall translation).

115. For further consideration of the questions considered here, the reader may consult the following: S. H. Nasr, *Man and Nature - The Spiritual Crisis of Modern Man* (Chicago: ABC International [Kazi Publications], 1998); idem, *The Need for a Sacred Science* (London: Curzon Press and Albany: State University of New York, 1993); idem, *Religion and the Order of Nature* (London and New York: Oxford University Press, 1996); S. C. Rockefeller and J. C. Elder, *Spirit and Nature* (Boston: Beacon Press, 1996); Roszak, *Where the Wasteland Ends*; Philip Sherrard, *The Rape of Man and Nature* (Ipswich: Golgonooza Press, 1987); M. E. Tucker and J. A. Grim, *Worldviews and Ecology* (Lewisburg PA: Bucknell University Press, 1993).

116. It originally appeared in *The Structurist*, Autumn Edition, 2000.

117. 'Recovering the Lost Millennium', *The Month*, February 1977, p. 58.

118. Ludwig von Bertalanffy, *Modern Theories of Development: An Introduction to Theoretical Biology*, trans. J.H. Woodger (New York: Harper Torchbook, 1962), p. 123.

119. Cited by von Bertalanffy, ibid., p. 12.

120. Eugene Odum, *Fundamentals of Ecology* (Philadelphia: W.B. Saunders, 1953), p. 5.

121. Henrick Kraemer, *The Christian Message in a Non-Christian World* (New York: Harper, 1938); cited in Robert T. Parsons, *Religion of an African Society* (Leiden: E. J. Brill, 1964), p. 176.

122. Gerardo Reichel-Dolmatoff, 'The Loom of Life: A Kogi Principle of Integration', *Journal of Latin American Lore* 4:1 (1978), pp. 5–27: pp. 23–4.

123. Ananda Coomaraswamy, *Symbolism in Indian Architecture* (Jaipur: The Historical Research Documentation Centre, 1983), p. 14. This book is a reprint of Coomaraswamy's article 'Symbolism of the Dome', originally published in the *Indian Historical Quarterly* 14 (1938), pp. 1–56.

124. *Midrash Tanhuma*, Pequde §3; cited in Raphael Patai, *Man and Temple in Ancient Jewish Myth and Ritual* (London: Thomas Nelson, 1947), p. 116.

125. *Bereshith Rabbati*, cited ibid., p. 114.

126. Titus Burckhardt, *Chartres and the Birth of the Cathedral*, trans. William Stoddart (Ipswich: Golgonooza Press, 1995), pp. 22ff.

127. Ibid, p. 17.

128. *Mishnah Sheqalim* 8.4–5; trans. Jacob Neusner, *The Mishnah: A New Translation* (New Haven and London: Yale University Press, 1988), p. 264. For discussion of this passage, and of the citations from Josephus below, see Margaret Barker, *The Gate of Heaven: The History and Symbolism of the Temple*

in Jerusalem (London: SPCK, 1991), pp. 106–7, 112–14.

129. Josephus, *Jewish War* 5.212–14; trans. H. St. J. Thackeray, *Josephus*, 9 vols (London: William Heinemann, 1928), iii.265.

130. Josephus, *Jewish Antiquities* 3.184; trans. Thackeray, *Josephus*, iv.405.

131. Reichel-Dolmatoff, ibid.

132. Walter F. Morris Jr, 'The Marketing of Maya Textiles in Highland Chiapas, Mexico', in Margot Blum Schevill, Janet Catherine Berlo and Edward B. Dwyer, eds, *Textile Traditions of Mesoamerica and the Andes: An Anthology* (New York and London: Garland, 1991), pp. 403–33: p. 404.

133. Natalie Tolbert, review of Danielle Geirnaert-Martin, 'The Woven Land of Laboya', *The Ecologist* 19.04.99 (unpublished).

134. Ibid.

135. Roger Caillois, *L'homme et le sacré* (Paris: Gallimard, 1988).

136. Fritz Schumacher, *Small is Beautiful: A Study of Economics as if People Mattered* (London: Abacus, 1974), pp. 84–96.

137. *The Dhammapada: The Sayings of the Buddha*, trans. Thomas Byrom (London: Wildwood House, 1976).

138. *The Dhammapada*, trans. Juan Mascaró (Harmondsworth: Penguin Books, 1973), p. 35.

139. See p. 119 above.

140. Marco Pallis, *The Way and the Mountain* (revised ed.; London: Peter Owen, 1991), p. 45.

141. The words are those of Dr James Watson, as reported in *The Daily Mail*, 17 May 2000.

142. HRH The Prince of Wales, 'A Reflection on the Reith Lectures', p.16.

143. Taken from the *Aṣṭasāhasrikā* and quoted in Anne Bancroft, ed., *The Buddha Speaks* (Boston: Shambhala, 2000), p. 117.

144. Stephen Batchelor, *Buddhism Without Beliefs* (New York: Riverhead Books, 1997), p. 77.

145. Fragment 12; and the citation of Heraclitus by Plato, Cratylus 402A. Texts with translation in G. S. Kirk and J. E. Raven, The Presocratic Philosophers (revised ed.; Cambridge: Cambridge University Press, 1963), pp. 196–7.

146. Majjhima Nikāya, Sutta 28.8; cited from Bhikkhu Ñanamoli and Bhikku Bodhi, trans., The Middle Length Discourses of the Buddha: A New Translation of the Majjhima Nikāya (Boston: Wisdom Publications, 1995), p. 283.

147. Ibid., Note 408, pp. 1231–2.

148. Joseph Campbell, The Hero with a Thousand Faces, Bollingen Series 17 (New York: Pantheon Books, 1949), pp. 49ff.

149. Ibid., p. 58.

150. Op. cit.; cf. note 2 above.

151. This is a paraphrase of the dedicatory verses to Nāgārjuna's treatise Mūlamadhyamakakārikā, taken from Stephen Batchelor, Verses From the Center: A Buddhist Vision of the Sublime (New York: Riverhead Books, 2000), p. 83.

152. This poem was first published in Resurgence 197 (1999), and has been reprinted in The Collected Poems of Kathleen Raine (Ipswich: Golgonooza Press, 2000), pp. 346–51.

Notes on Contributors

❧

WENDELL BERRY is a writer whose most recent publications include his *Selected Poems, Life is a Miracle*, and *Jayber Crow*. He lives and farms in his native Kentucky with his wife, Tanya. He is a Fellow of the Temenos Academy.

SUHEIL BUSHRUI is a distinguished author, poet, critic and translator. He presently holds the Bahá'í Chair for World Peace in the Center for International Development and Conflict Management at the University of Maryland. The foremost authority on the works of Kahlil Gibran, he is the author, with Joseph Jenkins, of the definitive biography *Kahlil Gibran, Man and Poet*. He is a Fellow of the Temenos Academy.

DAVID CADMAN was Chairman of the Trustees of the Prince's Foundation (1999–2001) and is a member of the Council of the Temenos Academy. He is a Visiting Professor at University College London and at the University of the West of England. He is the author of *The King Who Lost His Memory* and a number of tales and essays about our relationship with the land, Quakerism and Buddhism.

EDWARD GOLDSMITH is a Founder and Publisher of *The Ecologist*, the leading campaigning environmental magazine. He is the author of many articles and books including *The Way – An Ecological World-View*.

BRIAN GOODWIN is a Scholar in Residence at Schumacher College, Dartington, Devon, where he coordinates the MSc in Holistic Science. He was until recently Professor of Biology at the Open University. He is the author of *How the Leopard Changed its Spots*.

SATISH KUMAR, a former Jain Monk, is Director of Programmes at Schumacher College and editor of *Resurgence* magazine. His autobiography *No Destination* is published by Green Books.

SEYYED HOSSEIN NASR has been described as 'our greatest living philosopher'. University Professor of Islamic Studies, George Washington University, he is the author of many works including *The Need for a Sacred Science*, *Man and Nature* and *Religion and the Order of Nature*, three seminal books which explore the ecological wisdom to be found in the world's religions. He is a Fellow of the Temenos Academy.

JEREMY NAYDLER is a philosopher, gardener and cultural historian. He is the author of *Temple of the Cosmos*, *the Ancient Egyptian Experience of the Sacred* and *Goethe on Science*. He is the Founder/Director of The Jupiter Trust, an organization based in Oxford dedicated to developing and extending consciousness of the Sacred through regular lecture programmes and inspirational events.

KATHLEEN RAINE, C.B.E., the Founder of the Temenos Academy, is our most distinguished living poet and, through her advocacy of the primacy of the Imagination, a source of inspiration to many. Her *Collected Poems* was published in 2000; 2002 should see the re-issue of several of her out-of-print works on William Blake.

PHILIP SHERRARD (1922–1995) was, with Keith Critchlow, Brian Keeble and Kathleen Raine, one of the four co-founders of the review *Temenos*. Deeply influenced by the Traditionalist writers René Guénon, A. K. Coomaraswamy, Frithjof Schuon and others, and a member of the Greek Orthodox Church, Philip Sherrard's books include *The Rape of Man and Nature*, *The Sacred in Life and Art*, *Human Image, World Image*, and *Christianity: Lineaments of a Sacred Tradition*.

VANDANA SHIVA has been a visiting professor at several universities. She is a Founder Board Member of the International Forum on Globalisation and of the Women's Environment and Development Organisation, and she is Chair of the International Forum on Food and Agriculture. She has received numerous awards including the Alternative Nobel Prize. She is the author of *The Violence of the Green Revolution*, *Monocultures of the Mind*, *Most Farmers in India are Women* and *Tomorrow's Biodiversity*.